Reviews and critical articles covering the entire field of normal anatomy (cytology, histology, cyto- and histochemistry, electron microscopy, macroscopy, experimental morphology and embryology and comparative anatomy) are published in Advances in Anatomy, Embryology and Cell Biology. Papers dealing with anthropology and clinical morphology that aim to encourage cooperation between anatomy and related disciplines will also be accepted. Papers are normally commissioned. Original papers and communications may be submitted and will be considered for publication provided they meet the requirements of a review article and thus fit into the scope of "Advances". English language is preferred, but in exceptional cases French or German papers will be accepted.

It is a fundamental condition that submitted manuscripts have not been and will not simultaneously be submitted or published elsewhere. With the acceptance of a manuscript for publication, the publisher acquires full and exclusive copyright for all languages and countries.

Twenty-five copies of each paper are supplied free of charge.

Manuscripts should be addressed to

Prof. Dr. F. **BECK**, Howard Florey Institute, University of Melbourne, Parkville, 3000 Melbourne, Victoria, Australia

Prof. Dr. B. **CHRIST**, Anatomisches Institut der Universität Freiburg, Abteilung Anatomie II, Albertstr. 17, D-79104 Freiburg, Germany

Prof. Dr. W. **KRIZ,** Anatomisches Institut der Universität Heidelberg, Im Neuenheimer Feld 307, D-69120 Heidelberg, Germany

Prof. Dr. E. **MARANI**, Leiden University, Department of Physiology, Neuroregulation Group, P.O. Box 9604, 2300 RC Leiden, The Netherlands

Prof. Dr. R. **PUTZ**, Anatomische Anstalt der Universität München, Lehrstuhl Anatomie I, Pettenkoferstr. 11, D-80336 München, Germany

Prof. Dr. Dr. h.c. Y. **SANO,** Department of Anatomy, Kyoto Prefectural University of Medicine, Kawaramachi-Hirokoji, 602 Kyoto, Japan

Prof. Dr. Dr. h.c. T. H. **SCHIEBLER,** Anatomisches Institut der Universität, Koellikerstraße 6, D-97070 Würzburg, Germany

Prof. Dr. K. **ZILLES**, Universität Düsseldorf, Medizinische Einrichtungen, C. u. O. Vogt-Institut, Postfach 101007, D-40001 Düsseldorf, Germany

Advances in Anatomy
Embryology and Cell Biology

Vol. 155

Editors

F. Beck, Melbourne   B. Christ, Freiburg
W. Kriz, Heidelberg
E. Marani, Leiden   R. Putz, München
Y. Sano, Kyoto   T. H. Schiebler, Würzburg
K. Zilles, Düsseldorf

Springer-Verlag Berlin Heidelberg GmbH

A. Bozhilova-Pastirova   W. Ovtscharoff

The Inferior Oilvary Complex

With 39 Figures and 4 Tables

 Springer

ANASTASIA BOZHILOVA-PASTIROVA
WLADIMIR OVTSCHAROFF
Department of Anatomy and Histology
Medical University – Sofia
1. Sv. G. Sofiiski St.,
1431 Sofia, Bulgaria
e-mail: botzilov@medfac.acad.bg

ISSN 0301-5556
ISBN 978-3-540-67234-0

Library of Congress-Cataloging-in-Publication-Data

Die Deutsche Bibliothek - CIP-Einheitsaufnahme
Bozilova-Pastirova, Anastasia:
 The inferior Olivary complex / A. Bozhilova-Pastirova and W. Ovtscharoff.
- Berlin; Heidelberg; New York; Barcelona; Honkong; London; Milan; Paris;
Singapore; Tokyo: Springer, 2000
  (Advances in anatomy, embryology, and cell biology, Vol. 155)
  ISBN 978-3-540-67234-0   ISBN 978-3-642-57321-7 (eBook)
  DOI 10.1007/978-3-642-57321-7

© Springer-Verlag Berlin Heidelberg 2000
Originally published by Springer-Verlag Berlin Heidelberg New York in 2000

Production: PRO EDIT GmbH, 69126 Heidelberg, Germany
SPIN: 10718061   27/3136wg - 5 4 3 2 1 0

Contents

The Inferior Oivary Complex

1
Introduction

The inferior olivary complex is one of the precerebellar nuclei, which is marked by its close relation to the cerebellar molecular layer. Corpora olivares or olivary bodies was the first name for this nucleus that was given by Gabriele Falopio (Willis 1664). The term inferior olivary complex is used to refer to the nuclear mass comprising all subdivisions in this nucleus.

There are many opinions about the influence of the inferior olivary complex on the motor control function of the cerebellum (Marr 1969; Albus 1971; Llinás and Volkind 1973; Llinás 1974, 1989, 1991; Fujita 1982; Strata and Montarolo 1982; Ito 1990, 1993; Pellionisz and Llinás 1980; Strata 1984; McGaslin and Morgan 1987; Voogd 1989; Llinás and Welsh 1993; Jones et al. 1995; Keating and Thach 1995, 1997; Sugihara et al. 1995; Welsh et al. 1995, 1998; De Zeeuw and Koekkoek 1997; Lang et al. 1997; Rondi-Reig et al. 1997; De Zeeuw et al. 1998; Apps 1999; Apps and Lee 1999; Lang et al. 1999; Lui et al. 1999), but how this nuclear complex performs this role remains a matter of discussion.

The inferior olivary complex is the only source of climbing fibers to the cerebellar cortex (Desclin 1974), which exert synaptic influence on the Purkinje cells (Szentágothai and Rajkovits 1959, Eccles et al. 1966) making a one-to-one contact with them (Cajal 1911). Olivocerebellar climbing fibers are topographically organized (Armstrong 1974; Armstrong et al. 1974; Chan-Palay 1977; Chan-Palay et al. 1977; Groenewegen and Voogd 1977; McGrane et al. 1977; Groenewegen et al. 1979; Brodal 1980; Brodal and Kawamura 1980; Courville and Franco-Cantin 1980; Oscarsson 1980; Walberg 1980; Eisenman 1981; Brodal and Brodal 1981, 1982; Voogd 1982, 1989; Campbell and Armstrong 1983; Saigal et al. 1983; Whitworth et al. 1984; Wharton and Payne 1985; Bernard 1987; Buisseret-Delmas 1988a,b; Kanda et al. 1989; Takeda and Maekawa 1989; Apps 1990; Ruigrok et al. 1992; Buisseret-Delmas and Angaut 1993; Voogd et al. 1993; Tan et al. 1995; Apps 1998; Strata and Rossi 1998). These connections are selectively labeled with D-[3H]aspartate (Wiklund et al. 1982; 1984) and release a substance which is pharmacologically glutamate-like (Foster and Roberts 1983). Climbing fibers also send collaterals to the deep cerebellar nuclei (Van der Want and Voogd 1987; Van der Want et al. 1989; De Zeeuw et al. 1997b).

The olivary and corticonuclear fibers in the cerebellum are topographically organized by projection modules including olivary subnuclei, a cerebellar sagittal zone, and cerebellar nuclei (Voogd and Bigaré 1980). There is evidence that cerebellar nucleo-olivary and olivo-nuclear projections are reciprocal (Dietrichs et al. 1985; Dietrichs

1

and Walberg 1985, 1986, 1989; De Zeeuw et al. 1988; Van der Want et al. 1989; Ruigrok and Voogd 1990, 1995). Cerebellar nucleo-olivary projections to the contralateral inferior olivary complex (Walberg et al. 1962; Dom et al. 1973; Graybiel et al. 1973; Walberg 1974; Martin et al. 1976; Tolbert et al. 1976; Brown et al. 1977; Chan-Palay 1977; Saint-Cyr and Courville 1979; Angaut and Cicirata 1982; Courville et al. 1983a; De Zeeuw et al. 1988; Ikeda et al. 1989; Ruigrok and Voogd 1990, 1995) provide γ-aminobutyric acid (GABA)ergic feedback to the inferior olivary complex (Angaut and Sotelo 1987, 1989; Buisseret-Delmas et al. 1989; De Zeeuw et al. 1989a,b; Fredette and Mugnaini 1991; Ruigrok and Voogd 1990; De Zeeuw et al. 1997b; De Zeeuw and Koekkoek 1997; Ruigrok 1997; Lang et al. 1996; Barmack 1996; Barmack et al. 1998). Electrotonic coupling via dendro-dendritic gap junctions (Sotelo et al. 1974; Llinás et al. 1974; Gwyn et al. 1977; Rutherford and Gwyn 1977, 1980; Bozhilova and Ovtscharoff 1979; King 1980) appears to be essential for the relay and integration of information between inferior olivary neurons and synchronization of olivocerebellar activity (Llinás et al. 1974; Welsh et al. 1995) in different mammalian species. The electrotonic communication is modulated by GABAergic inputs from the cerebellar and vestibular nuclei (Nelson et al. 1984, 1989; Sotelo et al. 1986; Angaut and Sotelo 1987, 1989; Nelson and Mugnaini 1988, 1989; De Zeeuw et al. 1988, 1989a,b; 1990a,b,c; Ruigrok and Voogd 1990; Fredette and Mugnaini 1991; Barmark 1996; Lang et al. 1996; De Zeeuw et al. 1997b; De Zeeuw and Koekkoek 1997; Lang et al. 1997; Ruigrok 1997; Barmack et al. 1998). In addition, activation of the cerebellar nucleo-olivary connections may be influential in changing the state of coupling between olivary neurons (Llinás and Sasaki 1989; Ruigrok and Voogd 1995).

The dorsal accessory olive and caudal half of the medial accessory olive receive ascending somatosensory projections from the spinal cord, the dorsal column nuclei, the spinal trigeminal nucleus (Armstrong 1974; Boesten and Voogd 1975; Groenewegen et al. 1975; King et al. 1975; Cook and Wiesendanger 1976; Armstrong and Schild 1980; Buisseret-Delmas 1980; Courville et al. 1983b; Swenson and Castro 1983; Molinari 1984, 1985, 1987; Huerta et al. 1985; Alonso et al. 1986; Molinari and Starr 1989; Bull et al. 1990; Apps 1998), the lateral cervical nucleus (Berkley and Worden 1978; Molinari 1984), and the nucleus Z (Buisseret-Delmas and Bantini 1978). The spinal area of the inferior olivary complex receives projections from the reticular formation (Brown et al. 1977; Martin et al. 1977; Walberg 1982; Courville et al. 1983b; Bishop 1984). In most physiological studies it is accepted that spino-olivary neurons exert an excitatory effect on the olivary neurons in the spino-olivo-cerebellar circuitry (Oscarsson et al. 1969; Oscarsson 1980; Gellman et al. 1983) while the role of the medullar reticular neurons in the olivary circuitry is inhibitory (Llinás et al. 1974). Brain stem reticular nuclei are a source of serotoninergic afferents to the inferior olivary complex (Bishop and Ho 1986). Serotoninergic afferents to the inferior olivary complex (Sjölund et al. 1980; Wiklund et al. 1981a,b; Takeuchi and Sano 1983; King et al. 1984; Bishop and Ho 1986) are considered in the context of the effect of the serotonin modulation of inferior olivary oscillations and synchronicity (Sugihara et al. 1995). Finally, a spectrum of neuropeptides is demonstrated in axon terminals and varicosities of spinal and brain stem origins (for references, see Gregg and Bishop 1997). It is suggested (Gregg and Bishop 1997) that peptides could modulate the activity of the olivary neurons either by changes in firing rate or through interactions with other neurotransmitters.

The medial area of the medial accessory olive (dorsal cap, nucleus β, dorsomedial cell column, and ventrolateral outgrowth) receives afferent projections from pretectal, prepositus hypoglossi, parasolitary, and vestibular nuclei (Brodal and Torvik 1957; Mizuno et al 1973; Walberg 1974; Brown et al. 1977; Ito et al. 1978; Maekawa and Takeda 1979; Saint-Cyr and Courville 1980; Gerrits et al. 1985; Leonard et al. 1988; Nunes-Cardozo and Van der Want 1990; Van der Togt and Van der Want 1992; De Zeeuw et al. 1993; 1994; Balaban and Beryozkin 1994; Wentzel et al. 1995; Butter-Ennever et al. 1996). These visual-vestibular projections are GABAergic and cholinergic in rats, rabbits, cats, and monkeys (Horn and Hoffmann 1987; Nelson et al. 1986; Nelson and Mugnaini 1989; Fredette and Mugnaini 1981; De Zeeuw et al. 1993, 1994; Barmack et al 1998).

The rostral half of the medial accessory olive, dorsomedial cell column, and principal olive receive mesencephalo-diencephalic inputs (Walberg 1956; Saint-Cyr and Courville 1980, 1982; Saint-Cyr 1987; Spence and Saint-Gyr 1988; De Zeeuw et al. 1989a,b, 1990a; Ruigrok and Voogd 1995), which are excitatory (Jeneskog 1987). The mesencephalo-diencephalic region relays indirect projections from the prefrontal and motor cortex (Saint-Cyr 1983; Kitao et al. 1983; Nakamura et al. 1983; Onodera 1984) and direct cortico-olivary projections to the medial olive and the principal olive (Soasa Pinto and Brodal 1969).

Components of the inferior olivary complex are recognized in lower vertebrates up to mammals, including humans (Kooy 1917; Papez 1929; Ariens Kappers et al. 1936).

The inferior olivary neurons in fish project to the contralateral cerebellum (reviewed in Nieuwenhuys et al. 1998). Homologue cell groups of olivary neurons in amphibians and reptiles are the source of climbing fibers, which project to the contralateral cerebellar cortex (Bangma and ten Donkelaar 1982; Künzle and Wiklund 1982; Wilczynski 1982; Gonzalez et al. 1984; Künzle 1985; Van der Linde and ten Donkelaar 1987; Van der Linde et al. 1990).

Olivary neurons in birds send their climbing fibers to all parts of the contralateral cerebellar cortex (Whitlock 1952; Freedman et al. 1977). Climbing fibers in birds, which were described by Ramón y Cajal (1888), form synapses on the dendrites of Purkinje neurons (Mugnaini 1969). These connections are organized in a longitudinal zonal pattern (Clarke 1977; Freedman et al. 1977; Armstrong and Clarke 1979; Furber 1983; Arends and Voogd 1989; Feirabentd 1990; Lau et al. 1998). Ipsilaterally located olivocerebellar projection neurons in the chicken are reported but their significance remains unclear (Lopes-Raman and Armengol 1996).

The inferior olivary complex of mammals, including humans, consists of a well-delineated medial accessory, dorsal accessory, and a principal olive (Kooy 1917; Mareschal 1934; Ariens Kappers et al. 1936; Olszewski and Baxter 1954; Braak 1970; Bowman and Sladek 1973; Whitworth and Heines 1986a). These main subnuclei are present in a large number of mammals, but there are variations in the subnuclei among species (Korneliussen and Jansen 1964; Bowman and King 1973; Martin et al. 1975; Whitworth et al. 1983; 1984; Whitworth and Haines 1986b; reviewed in Whitworth and Haines 1986a).

On the basis of conventional histological methods and experimental studies of the olivocerebellar connections of the rat cerebellum, Azizi and Woodward (1987) demonstrated an alternative way of organizing the olivary cell groups into lamellae. They subdivided the medial accessory olive in horizontal, vertical, and rostral lamellae. The dorsal accessory olive and the principal olive are composed of ventral and dorsal

3

lamellae. According to Apps (1990), the olivary rostro-caudally oriented columns of the olivary neurons supplying the rat cerebellar cortex can be divided into smaller subgroups, and these olivary columns are comparable with the columnar organization of acetylcholinesterase activity in the rat, ferret, rabbit, and cat (Marani et al. 1977; Marani 1982).

All these observations can be linked to the efferent population of olivary neurons and paralleled to their morphological characteristics as they receive inputs from sets of the afferent system. Until now, topography, morphology, and connections of the inferior olivary complex have been studied mainly in the cat and more recently in the rat, opossum, and monkey, or there have been isolated investigations in other species (reviewed in Whitworth and Haines 1986a). However, phylogenetically, the inferior olivary complex reaches its greatest size and complexity in the human (Scheibel and Scheibel 1955).

The present study of the inferior olivary complex in several submammalian and mammalian species, including humans, is an attempt to provide new information and morphometric data about the normal structure of the olivary neurons and glial cells. Morphological data are correlated with electron-microscopical observations obtained from thin sections and freeze-etching replicas. Data about abnormal profiles in the neuropil and cells of the inferior complex are added and discussed as plastic changes.

2
Materials and Methods

For the morphological study of the inferior olivary nucleus the following vertebrate representatives were used: Fish, carp *(Cyprinus carpio)* (*n*=10); Amphibia, frogs *(Rana temporaria)* (*n*=10); Reptiles, lizards (*Lacerta muralis*) (*n*=10) and tortoises *(Tesdudo graeca)* (*n*=10); Birds, pigeons (*Columba livia*) (*n*=21). Mammals: Rodentia, rats (Sprague-Dawley and Wistar) (*n*=24) and ground squirrels (*Citellus citellus L.*) (*n*=34); Carnivora, cats (*Felis domestica*) (*n*=21); and 14 human brains stems.

2.1
Light Microscopy

Paraffin-embedded and Nissl-stained preparations were used for delineation of the boundaries of the inferior olivary complex and study of the neuronal cytology. We considered the following characteristics of the olivary neurons: the shape, size, and neuronal density in all areas of the inferior olivary complex. The study was carried out on the caudal portions of the brain stem containing the inferior olivary complex in five carp, five frogs, five lizards, five tortoises, five pigeons, five ground squirrels, three rats, and three cats. The animals were deeply anesthetized with ether or Thiopental (40 mg/kg) and fixed by immersion or transcardial perfusion with 10% formaldehyde. For Nissl staining after embedding in paraffin, the brains were cut transversally at 10-μm-thick sections and stained with cresyl violet.

The dendritic arborization of neurons in all subnuclei of the inferior olivary complex was studied in Golgi-stained material. The modification of the Golgi-Rio-Hortega method according to Stensaas and Stensaas (1968) proved useful for impregnating neurons and especially glial elements in low vertebrates. The brains of deeply anesthetized animals with ether or Thiopental (40 mg/kg) (five carp, five frogs, five lizards, five tortoises, five pigeons) were impregnated with this variation of the Golgi method. Five ground squirrels, three Sprague-Dawley rats, and three cats were treated according to the Golgi-Rio-Hortega impregnation as described by Sotelo and Palay (1968). Transcardial perfusion with 10% formaldehyde was used. The impregnated brains were embedded in paraffin and cut in serial transversal sections. Their thickness varied from 100 μm to 120 μm.

The inferior olivary complex in humans was examined using serial sections through the medulla oblongata of 14 children (about 1 year old) without neurological diseases (Table 1). This material was obtained according to the Bulgarian law within 10 h after death. The caudal portion of the brain stem containing the inferior olivary complex was dissected out and fixed by immersion in 10% formaldehyde. Four immature brain stems were embedded in paraffin and serial 10 μm transversely cut sections were stained with cresyl violet. The other ten human brain stems were used for Golgi analysis. They were prepared according to the modification of the Golgi-Rio-Hortega method as described by Sotelo and Palay (1968).

The internal structure of the inferior olivary complex was studied qualitatively and quantitatively in Nissl and Golgi preparations in paraffin series. The relative degree of basophilia was determined qualitatively. Morphometric analysis was performed using a microanalysis system (Olympus CUE-2) (primary magnification 20× and 40×). Data

Table 1. Summary of human brain stems without a history of neurological diseases used in this study

Number	Age in months	Sex Male (M)/ Female (F)	Pathological Diagnosis	Staining methods
1	2	M	Bronchiolitis	Golgi
2	1	M	Bronchiolitis	Golgi
3	2	F	Bronchopneumonia	Golgi
4	1	M	Tracheobronchitis	Golgi
5	3	M	Acute cardiovascular insufficiency	Golgi
6	5	F	Bronchiolitis	Nissl
7	5	F	Tracheobronchitis	Nissl
8	4	M	Tracheobronchitis	Golgi
9	6	F	Tracheobronchitis	Nissl
10	9	M	Asphyxia	Nissl
11	4	M	Bronchiolitis	Golgi
12	7	M	Asphyxia	Nissl
13	2	F	Bronchiolitis	Golgi
14	3	F	Bronchiolitis	Golgi

of the entire drawings were entered in the computer program (Olympus CUE-2), recorded automatically, and calculated. Statistical differences were examined by Student's t test. All values are presented as mean±standard error of the mean (SEM).

The shape of the inferior olivary complex throughout its length, the shape and size of neuronal areas, and the neuronal packing density were examined in Nissl-stained material. Analyses of neuronal packing densities were done for every tenth section of three brains from each species. The neurons were counted if an intact nucleolus was contained, in every tenth transversal sections on one side. The neuronal density was determined using the formula described by Abercrombie (1946) without correction because: first, the nucleoli are smaller than the section thickness (10 μm) and second, the total number of neurons in a particular area is equal with the number of nucleoli. Values of the shrinkage were taken into account (see Uylings et al. 1986).

Olivary neurons from carp, pigeons, ground squirrels, cats, and humans (n=100) were drawn. The cross-sectional areas, maximum and minimum diameters of neurons, which contained a nucleolus were measured. The ratio of its longest diameter and shortest diameter was computed.

The mean dendritic field area and its equivalent diameter in well-impregnated olivary neurons (n=5 per animal: pigeon, ground squirrel, cat, and 5 humans) were calculated by tracing the image of each cell. Two persons (A. Bozhilova-Pastirova and D. Brasizova) performed all measurements and any systematic error in the procedure may be assumed to be constant.

2.2
Electron Microscopy

All our observations were carried out using a Hitachi H-500 electron microscope. For the transmission electron microscopy three cats, five Wistar rats, ten ground squirrels in both periods of activity (the active summer period and winter sleep), and six pigeons were used. Animals of both sexes were anesthetized with sodium pentobarbital (40 mg/kg) and fixed by intracardial perfusion with 4% paraformaldehyde and 2% glutaraldehyde in 0.2 M cacodylate buffer, pH 7.2, containing 0.02% calcium chloride or 0.1 M phosphate buffer, pH 7.2. The perfused animal brains were stored overnight in the same fixative at 40 °C. The inferior olivary complex was dissected out and immersed in 1% OsO_4 with 0.2 M cacodylate buffer, pH 7.2, with 0.02% calcium chloride or 0.1 M phosphate buffer, pH 7.2, for 2 h at 4 C. After embedding in Durcupan ACM, ultrathin sections were cut with a Reichert or LKB ultramicrotome and were counterstained with uranyl acetate and lead citrate. Some tissue blocks were stained with uranyl acetate after osmification and others after buffer rinse (en block staining).

Two ground squirrels and two cats were fixed by intracardial perfusion with two solutions containing: 1.25% glutaraldehyde, 1% paraformaldehyde, 2% glutaraldehyde in 0.08 M cacodylate buffer, pH 7.2, containing 0.03% calcium chloride; and 4% paraformaldehyde, 5% glutaraldehyde in 0.08 M cacodylate buffer containing 0.03% calcium chloride, pH 7.2. The brains were stored overnight in the same fixative at 40 °C. Blocks containing the inferior olivary complex were treated for Golgi-electron microscopy according to Fairén et al. (1977). Deimpregnated sections were postfixed with 2% OsO_4, and embedded in Durcupan between transparent polypropylene sheets. The deimpregnated neurons were mounted on the polymerized Durcupan ACM blocks. Serial thin sections were cut with an LKB ultramicrotome, stained with uranyl acetate and lead citrate, and examined by electron microscope.

2.3
Freeze-etching

Adult cats ($n=2$), ground squirrels ($n=10$), Wister rats ($n=3$), and pigeons ($n=3$) were anesthetized with sodium pentobarbital (40 mg/kg, i. p.). Fixation was achieved by intracardial perfusion with 4% paraformaldehyde in 0.05 M phosphate buffer, pH 7.2. Tissue blocks containing the inferior olivary complex were immersed in 5% glutaraldehyde in 0.1 M phosphate buffer, pH 7.2, for 1 h at room temperature (Sandri et al. 1977), rinsed in phosphate buffer, pH 7.2, sucrose (2.72 mg in 40 ml 0.1 M phosphate buffer) and treated with buffered 25% glycerol. Tissue blocks were frozen in Freon 12, followed by liquid nitrogen. Freeze-etching was performed in Balzers BAF-400 D freeze-etching unit equipped with a thin quartz crystal to monitor the thickness of the Pt/C film. Successful replicas were examined by electron microscope.

Morphometric analysis was performed using a microanalysis system (Olympus CUE-2). Negative films of freeze-etched replicas (primary magnification ×24,000) were enlarged ten times. Statistical differences were examined by Student's t test.

2.4
Histochemistry

Ground squirrels in both periods of animal activity, active summer period ($n=2$) and winter sleep ($n=2$), were anesthetized with sodium pentobarbital (40 mg/kg, i. p.) and fixed by intracardial perfusion with 4% paraformaldehyde in 0.1 M phosphate buffer, pH 7.2. Brains were removed from the skull and kept in the same fixative for 5 h at 40° C. Tissue blocks comprising the inferior olivary complex were incubated for acid phosphatase reaction according to the lead method of Gomori (Gomori 1952). The specimens were prepared for electron microscopy using a routine procedure that included postfixing with 2% OsO_4 in 0.1 M phosphate buffer, pH 7.2, for 2 h at 40° C and embedding in Durcupan. Thin sections were cut with an LKB ultramicrotome and stained with lead citrate before examination by electron microscope.

2.5
Immunohistochemistry

Four adult cats, six male and six female Sprague-Dawley rats at postnatal day (P20) were used to study the immunoreactivity for GABA and parvalbumin. The day of birth was counted as P1. Animals were anesthetized with Thiopental (40 mg/kg, i.p.) and fixed by intracardial perfusion with 4% paraformaldehyde in 0.1 M phosphate buffer, pH 7.2. Brains were dissected free and kept in the same fixative for 1 h at 40° C, after which they were washed in phosphate-buffered saline (PBS), pH 7.4, in several changes and stored in the same buffer overnight at 40° C. Serial coronal sections throughout the extent of the inferior olivary complex were cut on a freezing micro-tome (Reichert-Jung) at 40 μm and every fifth section was taken for light-microscopical examination. For electron microscopy, 50-μm-thick sections were cut on a Vibratome. All sections were collected in PBS, pH 7.4, and then processed as free-floating sections.

Unspecific binding sites were inactivated by preincubation of sections with 5% normal goat serum for 1 h at room temperature. Sections were then incubated with antisera against GABA for 24–48 h at room temperature. Polyclonal anti-GABA (Sigma) antibodies (concentration of 1:4000) were prepared following the producer's description. Several washings in Tris buffered saline, pH 8.2, preceded the 1-h incubation in secondary antibody, goat anti-rabbit IgG, absorbed to 5 nm gold particles (Sigma), at a dilution of 1:100. After a series of rinsing, first in Tris buffered saline and then in distilled water, silver intensification with IntenSETMBL (Amersham) was performed. Thereafter, sections were stored in 2.5% sodium thiosulfate in distilled water for 2–3 min. The sections were coverslipped with Entellan for light-microscopical examination.

Other sections were produced for parvalbumin immunoreactivity using the avidin-biotin-peroxidase complex (ABC) method (Hsu et al. 1981). After brief PBS rinsing, pH 7.4, all sections were pretreated with 5% normal goat serum for 1 h. Sections were incubated for 1 h in mouse anti-parvalbumin serum (1:1000) and after several washings in PBS incubated with goat biotinylated antibody to mouse IgG (Vector Laboratories, 1:250) for 1 h. After rinsing in PBS these sections were incubated in ABC (Vector Laboratories) for 2 h, preincubated for 10 min in 0.05% diaminobenzidine tetrahydrochloride (DAB) in Tris-HCl buffer 0.05 M, pH 7.6. The incubation in 0.05% DAB and 0.01% H_2O_2 in Tris-HCl buffer 0.05 M, pH 7.6, lasted 10 min and was followed by rinsing in three changes of PBS. Sections were then dehydrated and mounted on slides and covered with Entellan for light microscopy. Olympus computer-assisted microanalysis system CUE2 was used to count immunoreactive neurons on serial sections of the inferior olivary complex.

Two cats, two ground squirrels, and two pigeon were used for immunostaining for glial fibrillary acidic protein (GFAP). Animals were anesthetized with Thiopental (40 mg/kg) and fixed by intracardial perfusion with 10% formaldehyde in 0.1 M phosphate buffer, pH 7.2. Brains were removed from the skull and kept in the same fixative for 1 h at 4 C, after which they were washed and embedded in paraffin. Serial 10 μm transversely cut sections throughout the extent of the inferior olivary complex were processed following the peroxidase-antiperoxidase (PAP) method of Sternberger (1979). Every fifth section was taken for light-microscopical examination and was mooted on glass slides, deparaffinized, and dehydrated. The immunohistochemical staining was performed using the Histo Gen immunoperoxidase stain kit, (Bio Genex Lab., Dublin, USA). After incubations with DAB and 0.01% H_2O_2 in Tris-HCl buffer 0.05 M, pH 7.6, for 10 min, sections were dehydrated and mounted on the glass slides covered with Entellan. Some sections were contrastained by cresyl violet.

3
Results

3.1
Light Microscopy

3.1.1
Topography of the Inferior Olivary Complex

The inferior olivary complex in Mammalia, including humans, is composed of three major subnuclei, namely, the medial accessory olive, the dorsal accessory olive, and the principal olive. These subnuclei can be subdivided into additional smaller subnuclei. The nuclear mass, throughout its whole length, lies dorsal to the pyramids and near the ventral border of the medulla oblongata.

The general topography of the inferior olivary complex in Rodentia is exemplified in this monograph by the description of the ground squirrel's inferior olivary complex. The description is carried out on transverse sections, which are Nissl-stained, and distances are calculated from the caudal to the rostral pole. The inferior olivary complex in adult animals measures between 3000 μm and 3100 μm.

The medial accessory olive is the first subdivision that starts at the caudal pole as the cell subgroup b (Fig. 1, I). At the subsequent levels in rostral direction, subgroup b is in communication with group a (Fig. 1, II) and group c (Fig. 1, III). Rostrally, these subgroups are mutually connected and subgroup a is more developed than other two subgroups. At these levels the dorsal cap also appears as a cluster of cells in close apposition with subgroup c (Fig. 1, IV), but these cell groups stay separated. Rostrally, the subgroup dorsal cap is fused with that of nucleus β (Fig. 1, V) and at the following levels is interconnected with other cell clusters of the medial accessory olive (Fig. 1, VI). At 1100 μm from the caudal pole, the ventrolateral outgrowth originates (Fig. 1, VII–IX). The next cell group of the medial accessory olive, namely, the dorsomedial cell column (Fig. 1, XI), appears approximately at level 2400 μm up to the most rostral levels of the inferior olivary complex.

The dorsal accessory olive appears as a single cell group at approximately 800–810 μm rostral to the caudal pole of the inferior olivary complex. This subgroup first lies dorsomedially to the medial accessory olive (Fig. 1, V–VI), and before the appearance of the principal olive at approximately the middle of the nuclear complex extends laterally (Fig. 1, VII). The dorsal accessory olive is in continuity with the dorsal lamella at approximately 1700 μm rostral to the caudal pole (Fig. 1, IX) and ventral lamella of the principal olive (at 2300–2400 μm) (Fig. 1, XI–XII), and both subnuclei reach the rostral pole of the inferior olivary complex (Fig. 1, XIV).

The principal olive appears in the rostral direction as a cell group located between both accessory olives at about the middle of the inferior olivary complex (Fig. 1, VIII). In the rostral direction, ventral and dorsal lamellae are in contact and form the hilus, which continues in a dorsomedial direction (Fig. 1, XI–X). Rostrally, the dorsal lamella (Fig. 1, XI) and the ventral lamella (Fig. 1, XI–XII) are joined with the dorsal accessory olive. The ventral cell group of the principal olive presents the rostral pole in continuity with the dorsal cell group of the dorsal accessory olive (Fig. 1, XIV).

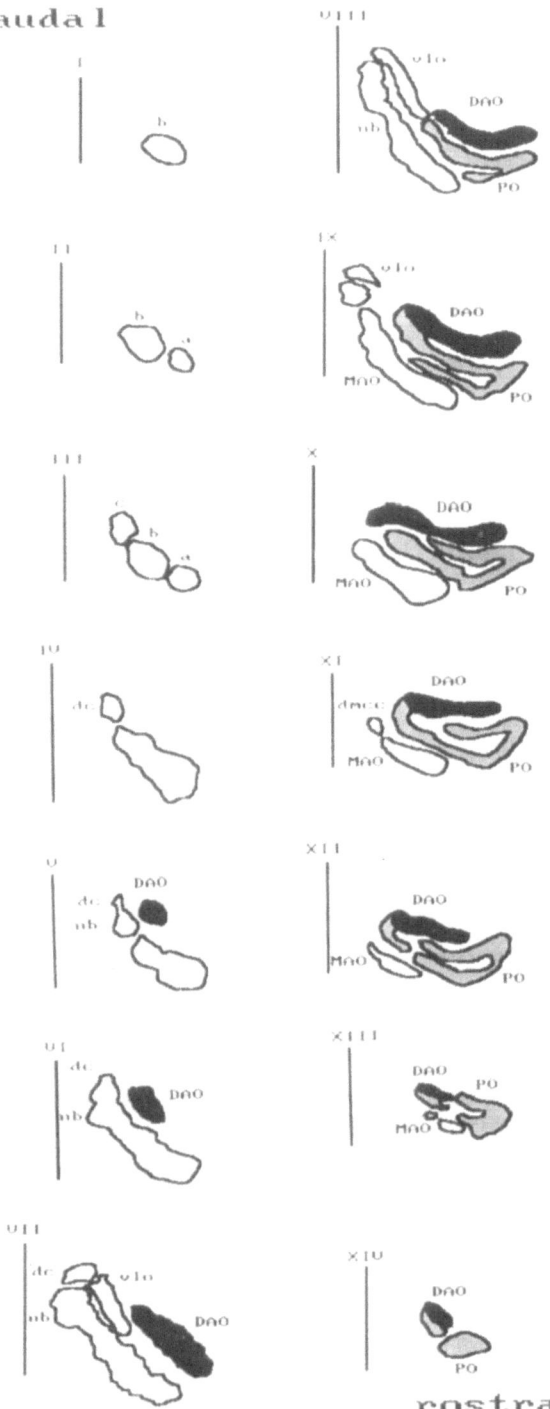

Fig. 1. Image analyzing system Olympus CUE-2 drawing of transversal Nissl-stained sections through the ground squirrel inferior olivary complex: subgroups *a*, *b*, *c* of the caudal medial accessory olive (*MAO*), dorsal cap (*dc*), nucleus β (*nb*); ventrolateral outgrowth (*vlo*), dorsal accessory olive (*DAO*), and principal olive (*PO*). ×20

caudal

rostral

10

The rat inferior olivary complex, as in the ground squirrel, contains a medial accessory olive, a dorsal accessory olive, and a principal olive, and the parcellation in our Nissl-stained material confirms the analysis of Gwyn et al. (1977).

The inferior olivary complex in the cat and human is subdivided into the same three components as in other mammalian species. In human inferior olivary complex, the most developed subnucleus is the principal olive. It is characterized by extensive infoldings along the ventral and dorsal lamellae and along its hilus (see Kooy 1917).

The inferior olivary complex of submammalian vertebrates forms a column of cells in a corresponding position in the rhombencephalon as the homologue of the mammalian inferior olivary complex.   .

The inferior olivary complex in carp is the most ventro-laterally extending nucleus in the caudal one-half of the rhombencephalon and is bordered medially by the raphe. This nuclear complex is irregular in form and reaches its largest extent in its middle third, tapering at the more caudal and rostral levels. It consists of a group of neurons, which may be divided into a dorsal and ventral part. Dorsally, a narrow cell-sparse zone separates the neurons of this accumulation from ventrally located neurons.

A small group of neurons with comparable, if not identical, characteristics of the olivary neurons of the carp is observed in the sections of the caudal one-half of the rhombencephalon in the frog, lizard, and tortoise. These neurons are scattered throughout the ventral parts of the caudal brain stem and are placed more laterally to the raphe. At the caudal levels of the rhombencephalon of the tortoise such types of neurons are sparsely present. These cell groups can still be recognized because they are surrounded by dispersed large reticular neurons.

The inferior olivary complex in the pigeon is located near the ventral borders in the caudal portion of the medulla oblongata, medial to the raphe and lateral to the reticular nucleus. This cell column extends rostro-caudally for a distance ranging between 1.6 to 1.75 mm. The inferior olivary complex is divided into a ventral and a dorsal lamella. The bundle of hypoglossal fibers passes through the lateral field of the inferior olivary complex. At the caudal pole, the inferior olivary complex only consists of a dorsal lamella, medial to the raphe. In the rostral direction, the dorsal lamella begins to extend more medially. Near the midline a number of ventral lamella neurons form a distinct group, ventral and lateral to the dorsal lamella. In the rostral direction the ventral lamella tends to sweep up medially to merge with the dorsal lamella. A lateral directed hilus of the U-shaped configuration exist only for a short distance. More rostrally, the cell column exhibits no division into a dorsal and a ventral lamella. The lamella ends as small clusters of neurons.

The inferior olivary complex in all species examined, including humans, contains rounded to oval-shaped neurons and tends to be arranged in clusters between groups of myelinated fibers. The olivary neurons possess a diffuse basophilia and a relatively large nucleus.

Besides the olivary neurons just described, single reticular neurons could be seen in close apposition to the boundary of both olivary lamellae in the pigeon. Such reticular-like neurons are located either around the subnuclei or at the boundary in the inferior olivary complex of the ground squirrel, rat, and cat. Occasionally, these neurons can be encountered as small groups near the pyramids. Reticular-like neurons are distinctly larger than typical olivary neurons. In humans, some neurons in the caudal parts of the medial accessory olive indeed have the same morphological characteristics as the reticular neurons in other species.

Statistical analyses of Nissl-stained material included measurements of the neuronal mean area and its mean maximal and minimal diameters in the carp, pigeon, ground squirrel, cat, and human inferior olivary complex (Fig. 2), which are summarized in the frequency histograms (Figs. 3, 4). There is an apparent increase in the mean area, and maximal and minimal diameters of the olivary neurons from the carp to the cat with the largest size being reached in the human (Fig. 2).

The neurons in the submammalian inferior olivary complex range in area between 50 and 400 μm^2 (Fig. 3) and in maximal and minimal diameter (Fig. 4). In the carp, smaller neurons contribute 62% of the population and larger neurons 38%, giving a

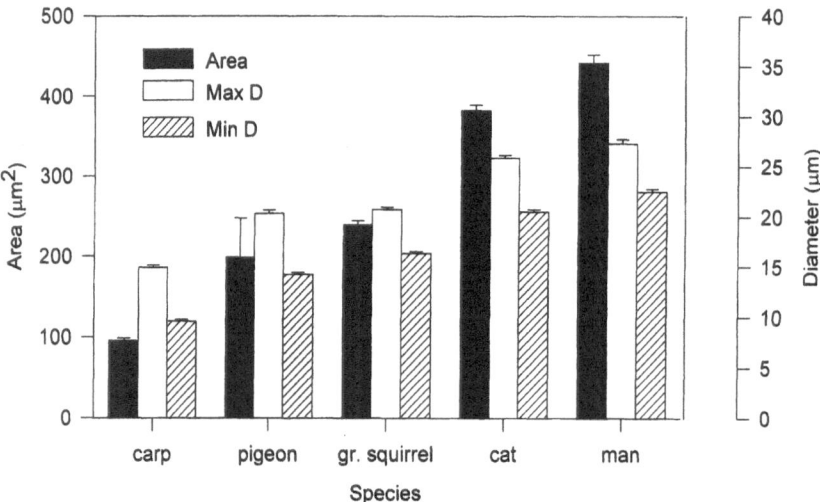

Fig. 2. Summary of neuronal mean areas (μm^2), and mean maximal and minimal diameters (μm) of Nissl-stained neurons ($n=100$) in the carp, pigeon, ground squirrel, cat, and human inferior olivary complex. Values are expressed as mean±SEM

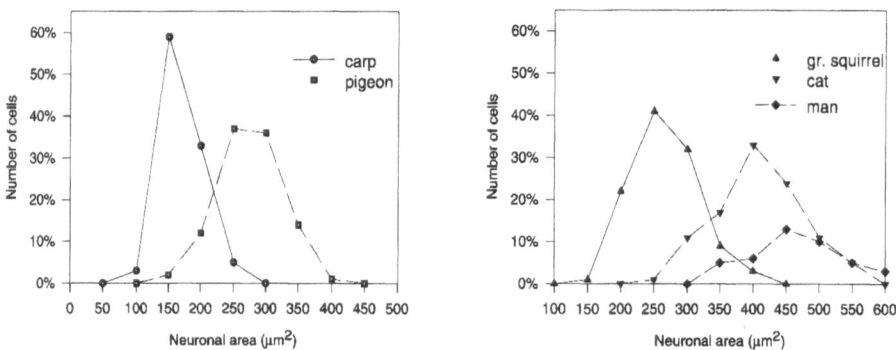

Fig. 3. Frequency histograms illustrating the unimodal distribution of the area of olivary neurons in the carp, pigeon, ground squirrel, cat, and human

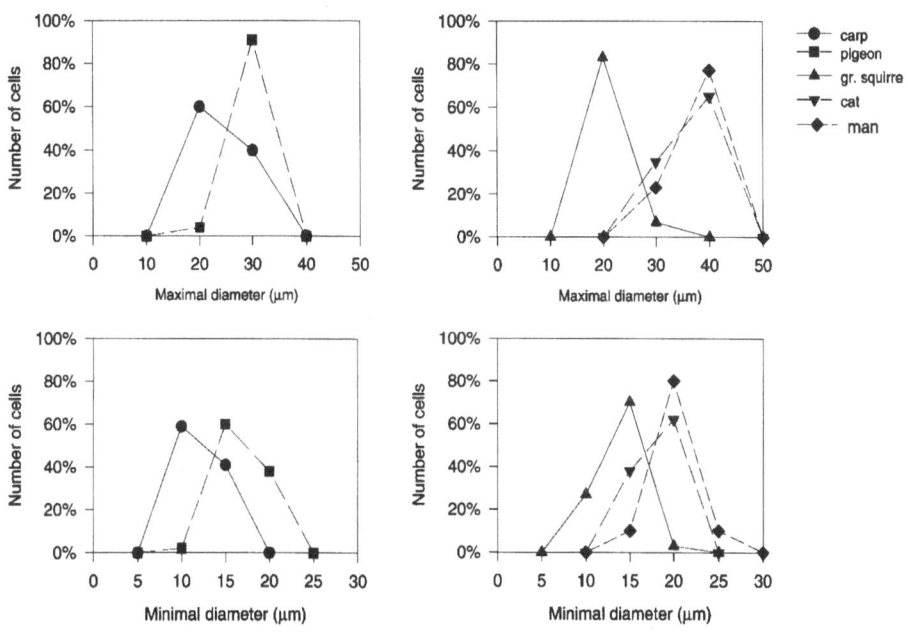

Fig. 4. Frequency histograms comparing the minimal and maximal diameters of the olivary neurons in the carp, pigeon, ground squirrel, cat, and human

ratio of 1.6:1 between small and large cells. The perikarya in the pigeon olivary complex show a range in area between 50 and 400 μm^2 and form a ratio of 1:1.

The olivary neurons in mammalian species, including humans, range in area between 100 and 600 μm^2 (Fig. 3) with a maximal diameter between 10 and 50 μm or a minimal diameter between 5 and 30 μm (Fig. 4). The ground squirrel olivary neurons range in area between 150 and 450 μm^2, giving a ratio of 0.2:1 between the smaller (with a break at 200 μm^2) and the set of larger neurons. In the cat, the inferior olivary complex cross-sectional area ranges between 200 and 600 μm^2 (with a break at 250 μm^2), and this ratio is 0.4:1. The human olivary neurons range in area between 300 and 600 μm^2 (with a break at 250 μm^2), giving a ratio of 0.3:1 between small and large neurons.

Based on these ratios between the smaller and larger neurons the submammalian inferior olivary complexes are comparable (1.6:1 in the carp and 1:1 in the pigeon). The mammalian inferior olivary complex can be grouped together, too. These ratios within the mammalians are comparable (0.2:1 in the ground squirrel, 0.3:1 in the cat, and 0.4:1 in the human), but differ from those in the submammalian inferior olivary complex. Moreover, neurons are more uniform in size in the mammalian inferior olivary complex.

The shape of the neuronal perikarya can be represented by the elongation index, which is the ratio of the maximal diameter to the minimal diameter. Most of the olivary neurons in submammalian species have slightly asymmetrical cell bodies with a ratio greater than 1:1.2 (elongated index 1:1.5 in the carp and 1:1.4 in the pigeon inferior olivary complex). Almost rounded cell bodies are present in the inferior olivary complex of the ground squirrel, cat and human (elongated index, 1:1.2).

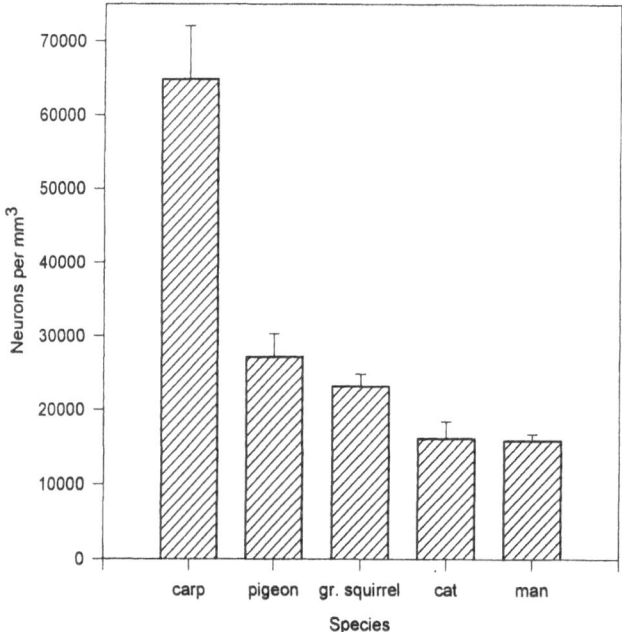

Fig. 5. Number of Nissl-stained olivary neurons per mm3 in the carp, pigeon, ground squirrel, cat, and human. Values are presented as mean±SEM

Measurements of the mean neuronal density per mm3 through the inferior olivary complex for all species demonstrate a decrease in number of neurons per mm3 towards humans (Fig. 5). The results on the soma size unequivocally show an increase from carp to humans (Fig. 2). Thus, calculating from our serial sections, the lowest number of neurons per mm3 were found in the human, but the neurons have the largest soma size.

3.1.2
Inferior Olivary Complex in Golgi Preparations

Golgi-impregnated olivary neurons examined in all specimens, including human, are round to oval-shaped with dendrites which extend in all directions. Several cell types can be discerned according to their patterns of dendritic arborization. Dendrites of the first type of olivary neuron radiate straight from the perikaryon. Many dendrites of the second neuronal type follow a curved part around the cell body and have a ball-like appearance. The reticular type of neuron with a marginal position or lying among the typical olivary neurons can be noted as an additional cell type.

3.1.2.1
Dendritic Morphology of Submammalian Inferior Olivary Neurons
Only the first radiating cell type exists in the inferior olivary complex of the carp (Fig. 6a). The primary dendrites branch once or twice to form secondary and tertiary

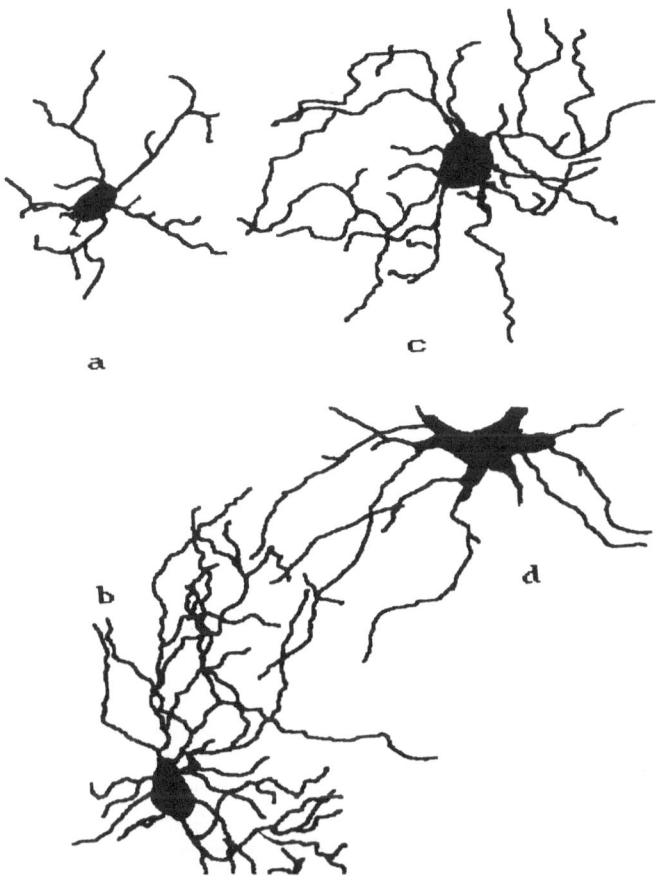

Fig. 6a-d. Image analyzing system Olympus CUE-2 drawing of Golgi-impregnated neurons in the carp and pigeon inferior olivary complex. **a** First radial type olivary neuron in the carp. **b** First-type olivary neuron in the pigeon. **c** Second-type olivary neuron in the pigeon. **d** A reticular neuron sends its dendrites into the dorsal lamella of the pigeon inferior olivary complex. ×20

dendrites. Dendrites of these olivary neurons carry few spines possessing a wide variety of sizes and shapes.

Neurons with dendritic morphology comparable to the carp olivary neurons can be distinguished in the caudal one-half of the rhombencephalon of the frog, lizard, and tortoise. Other types are absent.

The largest population of well-impregnated olivary neurons in the pigeon belongs to the radiated type (Fig. 6b). Usually four to five thin primary dendrites (about 2–3 μm in diameter) originate from either pole of the ovoid cell body and branch to form secondary and tertiary dendrites or remain unbranched before ending. Their radial dendrites produce an elliptical to spherical dendrite field and remain within the boundary of the inferior olivary lamellae. The dendrites of most olivary neurons in the pigeon exhibit only a few large spines that are scattered over all regions of the dendrites. Only occasionally were small spines observed on the cell body.

Dendrites of some olivary neurons follow a slightly curved path forming gentle curves. They are characterized by a ball-like arrangement (Fig. 6c). These second ball-like types of olivary neurons have an oval to rounded dendritic field. Their dendrites are relatively aspiny except for a few short spines at their distal dendrites. Only occasionally were small spines observed on the cell body. This second type of olivary neuron is located in the intermediate parts of the dorsal lamella at its rostral levels.

Golgi-impregnated large reticular neurons at the lamella boundary of the inferior olivary complex in the pigeon are multipolar or fusiform. Their dendrites are radially oriented with numerous short and long spines. They extend into the lamellae region of the inferior olivary complex (Fig. 6d). These neurons possess four to seven primary dendrites ranging in diameter. Most primary dendrites branch once or twice before ending, but some remain unbranched.

3.1.2.2
Dendritic Patterns of Mammalian Olivary Neurons
The perikarya of Golgi-impregnated neurons in the inferior olivary complex of the ground squirrel, rat, cat, and human are oval to round in shape and occasionally their somata are in close apposition. The dendrites of most olivary neurons have mainly short and stalked spines that are sparsely distributed on the primary, secondary, and tertiary dendrites. Many of the spines are complex racemose formations, the so-called spine-crowned appendages. These formations are located along most regions of the dendrites and neuronal cell body. Some neurons in the cat dorsal accessory olivary nucleus exhibit secondary and tertiary dendrites with swelling or varicosities. They are irregularly spaced and linked by thin segments. Dendrites remain within the confines of the subnuclei and partially overlap the dendrites of adjacent neurons. In the ground squirrel, rat, and cat inferior olivary complex there are examples of over-lapping with the dendrites of neighboring subnuclei. Terminal dendrites of some neurons in the ground squirrel and rat medial accessory olive lie in close vicinity to the middle of the medulla oblongata and are suspected to pass to the contralateral subnucleus.

On the basis of the dendritic pattern, three types of neurons are found in the mammalian inferior olivary complex, including humans. The first type of olivary neuron with radial arranged dendrites is predominantly present in caudal portions of the medial accessory olive and in the dorsal olive. This type is very rarely found in subnuclei of the human inferior olivary complex. The second neuron type, having a ball-like appearance, is predominantly present in the principal olive and the rostral half of the medial accessory olive. The third neuron type, in marginal position with dendrites projecting into the nucleus, is situated at all subnuclei. An additional cell type is the reticular-like type of neuron with a peripheral position or lying among the typical olivary neurons. In the human inferior olivary complex some neurons in the medial accessory olive possess themorphological characteristics of reticular neurons.

The first or radial type of well-impregnated olivary neuron has an oval to rounded form. This neuron type possesses three to five primary dendrites, which branch once or twice to form secondary and tertiary dendrites. The radial oriented dendrites form an elliptical or spherical dendritic field (Fig. 7a).

The second or ball-like type of olivary neuron contains either an oval or stellate form with six or more primary dendrites which branch to form secondary and

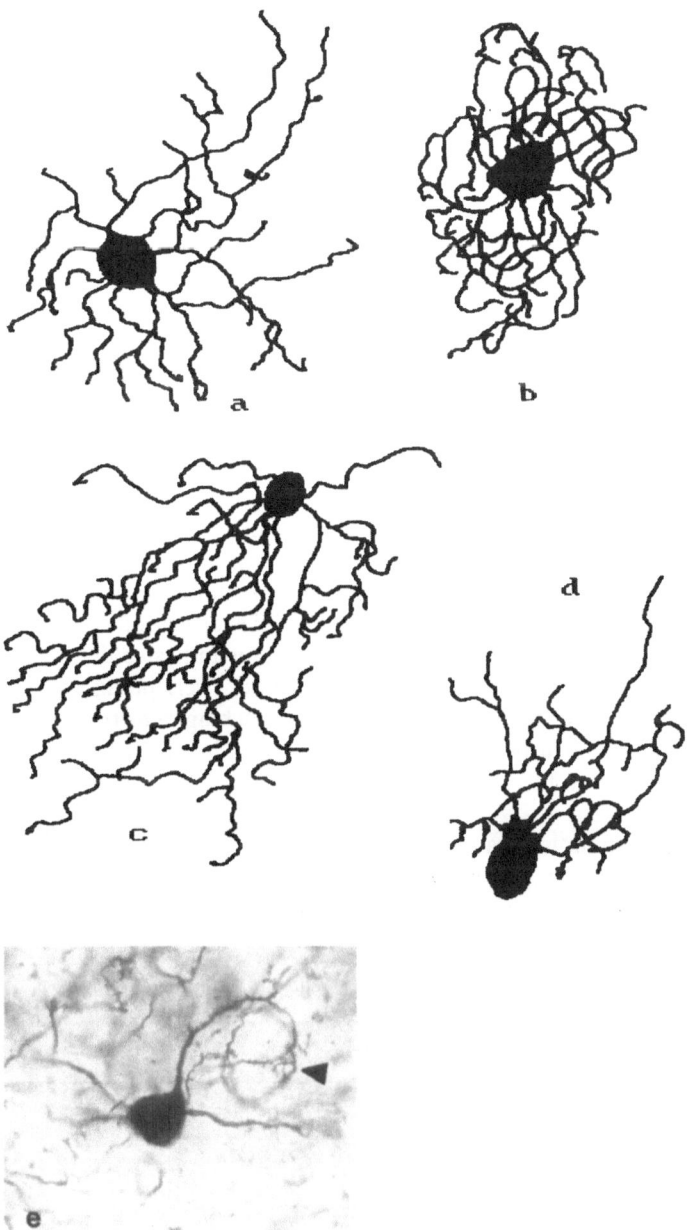

Fig. 7a-e. Image analyzing system Olympus CUE-2 drawing and a photograph of the Golgi-impregnated neurons in the mammalian inferior olivary complex. **a** First-type olivary neuron in the cat. **b** Second-type olivary neuron in humans **c, d** Third-type olivary neuron in the ground squirrel and human, respectively. ×20. **e** Light micrograph of a third-type olivary neuron with gentle bends (*arrow*) along the length of secondary and tertiary dendrites in the ground squirrel. ×157

17

tertiary dendrites. The primary dendrites are radially oriented and branch once or twice. The secondary and tertiary dendrites follow a curved path giving an appearance of coiled ball (Fig. 7b).

The third type of olivary neuron is located along the borders of the inferior olivary complex, especially in the lamellae of the principal olive. The dendrites of these marginal neurons are primarily oriented towards the center of the olivary lamella. Thus, the region of the cell body, which delimits the lamellae border, is devoid of dendrites (Fig. 7c,d). Occasionally, gentle bends are formed along the length of secondary and tertiary dendrites (Fig. 7e).

The reticular neurons can be located either in or around the inferior olivary complex of the ground squirrel, rat, and cat. These neurons are polygonal or fusiform and medium to large in size. The dendrites of Golgi-impregnated reticular neurons overlap with those of adjacent neurons at the boundaries of the inferior olivary complex. The dendrites of this neuronal type exhibit only long and short scattered spines.

Morphometric data of the mean area of the dendrite field and its equivalent diameter are summarized in Table 2. Measurements of mean dendritic field area in the inferior olivary complex for all mammalian species demonstrate a decrease in size. Mean dendritic field areas generally vary from one type to another providing one criterion for subdivision. The results on the packing densities of neurons per mm3 unequivocally show a decrease towards humans (Fig. 5). This indicates that in the submammalian inferior olivary complex, dendrites are more overlapping and the greater dendritic field area is concomitant with a greater packing density. Thus, in humans, the lowest number of neurons is present per mm3 with the smallest dendritic field area. Reticular neurons in the cat, rat, and ground squirrel possess the largest dendritic field area.

In our Golgi-Rio-Hortega-impregnated material dendrites are mainly stained but axons of some olivary neurons can be followed for a long distance in the contralateral direction before leaving the plan of section. These processes are smooth and the axon

Table 2. Histograms of type I, II, and III Golgi-stained neurons in the inferior olivary complex

	Type I	Type II	Type III
In the submammalian inferior olivary complex			
Carp	5468±408		
Pigeon	15566±2175	7031±57	
In the mammalian inferior olivary complex			
Ground squirrel	14050±123	13641.8±159	17579±124
Cat	13210.7±147	11891.3±335	13967±402
Human	5920±171	3321±66	7125±584

The mean dendritic field area (DF) given in μm^2. Values are expressed as mean±SEM.

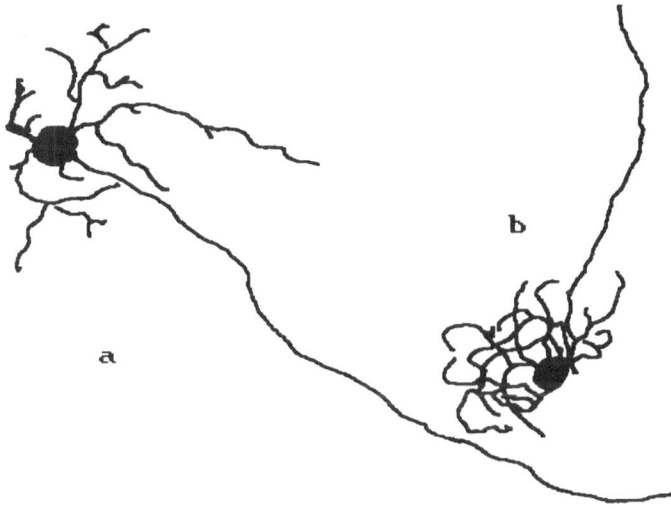

Fig. 8. Image analyzing system Olympus CUE-2 drawing of Golgi-impregnated olivary neurons. **a** An axon arises from the soma in the cat. **b** An axon arises from the proximal dendrite in the ground squirrel. ×20

portions visualized in our material do not form collaterals. The axons of the olivary neurons originate from the neuronal somata (Fig. 8a) as well as from the proximal dendrite (Fig. 8b).

3.1.3
GABA and Parvalbumin Immunoreactivity in the Inferior Olivary Complex

The GABAimmunostaining method provides information on the architecture of the three main subnuclei, associated smaller subnuclei, and on the spatial distribution pattern of the immunoreactive structures in the cat inferior olivary complex. It allows a selective study of the inferior olivary complex because the interlamellar zones and regions around this nuclear complex have a lower immunoreactivity. The cat inferior olivary complex contains a high density of GABA-immunoreactive punctuate structures representing axon terminals and varicosities. They are localized either among elements in the neuropil or close to the perikaryon, thus giving an impression of the boundaries of the entire cell. GABA-immunoreactive punctuate structures vary in size and density in the different subnuclei but also in the different parts of the territory of every subnucleus. Quantitative studies show that the average density of GABA-immunoreactive terminals per mm2 is highest in the nucleus β, while in the portions of the rostral medial accessory olive it is the lowest (Table 3).

GABA-immunoreactive neurons made up a heterogeneous population of neurons and displayed a variety of sizes (10–20 µm in diameter) and various distribution patterns in the cat inferior olivary complex. Only a few GABA-immunoreactive neurons are scattered in the cat medial and dorsal accessory olives. Larger GABA-im-

Table 3. The number of GABA-immunoreactive axon terminals per μm^2 in the nucleus β, dorsal accessory olivary nucleus (DAO), caudal medial accessory olive (MAOc), dorsal cap of Kooy (dc), principal olive (PO), and rostral medial accessory olive (MAOr) in the cat inferior olivary complex

	Number of axon terminals
Nucleus β	37,666±4115
DAO	31,567±4244
MAOc	27,933±1338
dc	24,000±3724
PO	22,633±2631
MAOr	19,657±2986

Values are presented as mean±SEM.

Table 4. The number of parvalbumin- (PA) and GABA-immunoreactive (GABA) olivary neurons per μm^2 in male and female rats at postnatal day 20

	PA	GABA
Male (M)	75±8	199±12
Female (F)	152±22	305±20

Values are presented as mean±SEM. PA (M) and PA (F) values are significantly different ($t=3295$, $p<0.01$). The GABA (M) value is significantly different from GABA (F) ($t=4619$, $p<0.01$).

munoreactive neurons are located in a periolivary or interolivary position and were identified as reticular-like neurons. These neurons are multipolar with a rounded soma up to 20 μm in diameter. It is interesting to note that small to medium-sized parvalbumin-immunoreactive neurons were observed in an intraolivary position and larger ones in peri- or intraolivary positions (A. Bozhilova-Pastirova, unpublished data).

In the rat inferior olivary complex, GABA-immunoreactive perikarya were observed at postnatal day 20. These perikarya resided on the dorsal cap and some small parts of the nucleus β. A few scattered parvalbumin-immunoreactive neurons were also found in all of the above mentioned areas. In addition, the numbers of GABA- and parvalbumin-immunoreactive neurons are different in male and female rats at postnatal day 20. The quantitative data obtained from these animals are summarized in Table 4. The average density of the parvalbumin-immunoreactive olivary neurons per mm2 for 20-day-old female rats is greater than for males. There is a statistically significant decrease in neuronal density from females to males ($t=3.295$, $p<0.05$).

In the female inferior olivary complex there is an increase in the average density of GABA-immunoreactive olivary neurons per mm2 compared to the male on the same postnatal day (Table 4). These sex differences are statistically significant ($t=3.295$, $p<0.01$).

The ratios between density of GABA- and parvalbumin-immunoreactive olivary neurons for male and female rats are comparable (2.0066:1 in females and 2.66:1 in males). Moreover, the calculated parvalbumin-immunoreactive olivary neurons appear to be 49.7% of the GABA-immunoreactive neurons in females and 37.5% in males.

3.2
Electron Microscopy

The fine structure of neurons and glial cells in the inferior olivary complex was studied in pigeons, rats, ground squirrels, and cats. Ultrastructural characteristics from thin sections and freeze-etched replicas are included in this chapter.

3.2.1
Neuronal Somata

For the classification of the olivary neurons, the size of the cross-sectioned area of the perikarya and nucleus is used. Medium-sized neurons (20–25 μm) are the predominate type. This type of olivary neuron is found in all the animals examined and it consists of round to oval or polygonal perikarya.

All neurons in the inferior olivary complex in every species and all the animals examined are characterized by an eccentrically located nucleus. Nuclei can be spherical to oval and smooth, or irregular and indented. Many olivary neurons have one or more invaginations of the nuclear envelope. The nucleus contains a dense and spheroid nucleolus with fibrillar and granular components.

The cytoplasm of olivary neurons contains a moderately high concentration of mitochondria, free ribosomes, and ribosomal clusters. The rough endoplasmic reticulum consists of parallel arrays and a few loosely arranged cisterns. Subsurface cisterns are common. Golgi complexes are well developed and spread from around the nucleus to the base of dendrites. Multivesicular bodies, lysosomes, and lipofuscin granules are distributed in the cytoplasm. Olivary neurons in the pigeon contain very few lysosomes and almost no lipofuscin granules. Neurofilaments and neurotubules are loosely arranged in the perikarya, but are organized as parallel bundles in proximal dendrites.

The plasmalemma of the olivary neurons is for the most part covered by astrocytes (Fig. 9) or oligodendrocytes and their processes. Most of the olivary neurons are aggregated in groups (Fig. 10). Plasma membranes of some adjacent perikarya are frequently closely apposed. Membrane appositions are also present between perikarya and dendrites up to 20 μm in length. Occasionally, they may be reinforced by puncta adhaerentia. Some neurons, predominately in the caudal half of the medial accessory olive, exhibit cilia arising near the Golgi apparatus. They have an axoneme with nine peripheral microtubules or are arranged as 8+1 (Fig. 11).

All neurons in the inferior olivary complex have few axosomatic synapses. Two types of axon terminals synapse on the spine-like protrusions or other regions of the neuronal plasma membrane. One type contains predominantly round vesicles and forms asymmetric synaptic junctions. Such axon terminals are, however, rarely found

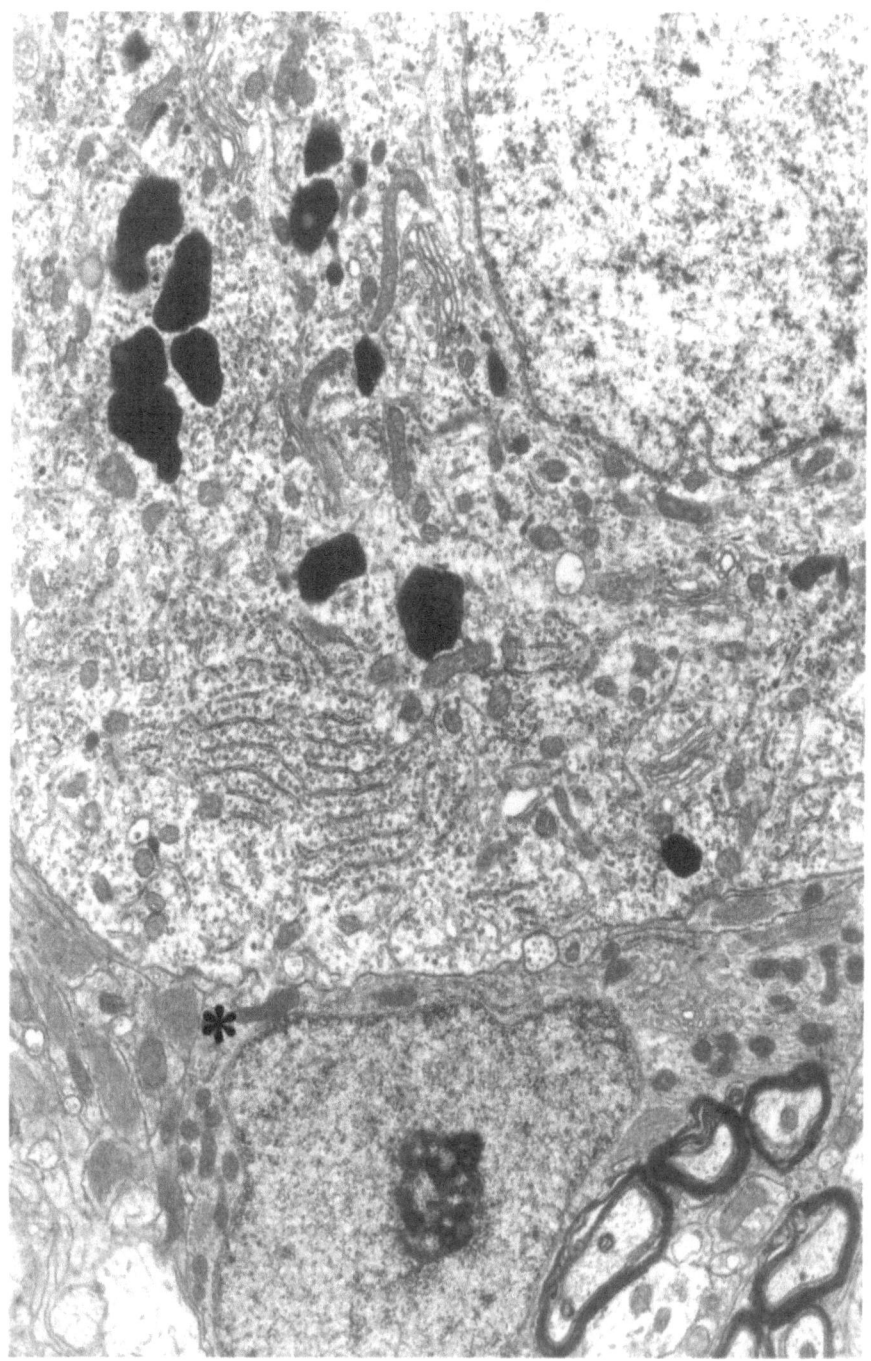

Fig. 9. Electron micrograph of an inferior olivary neuron with a satellite astrocyte (*asterisk*). Cat. ×11,500

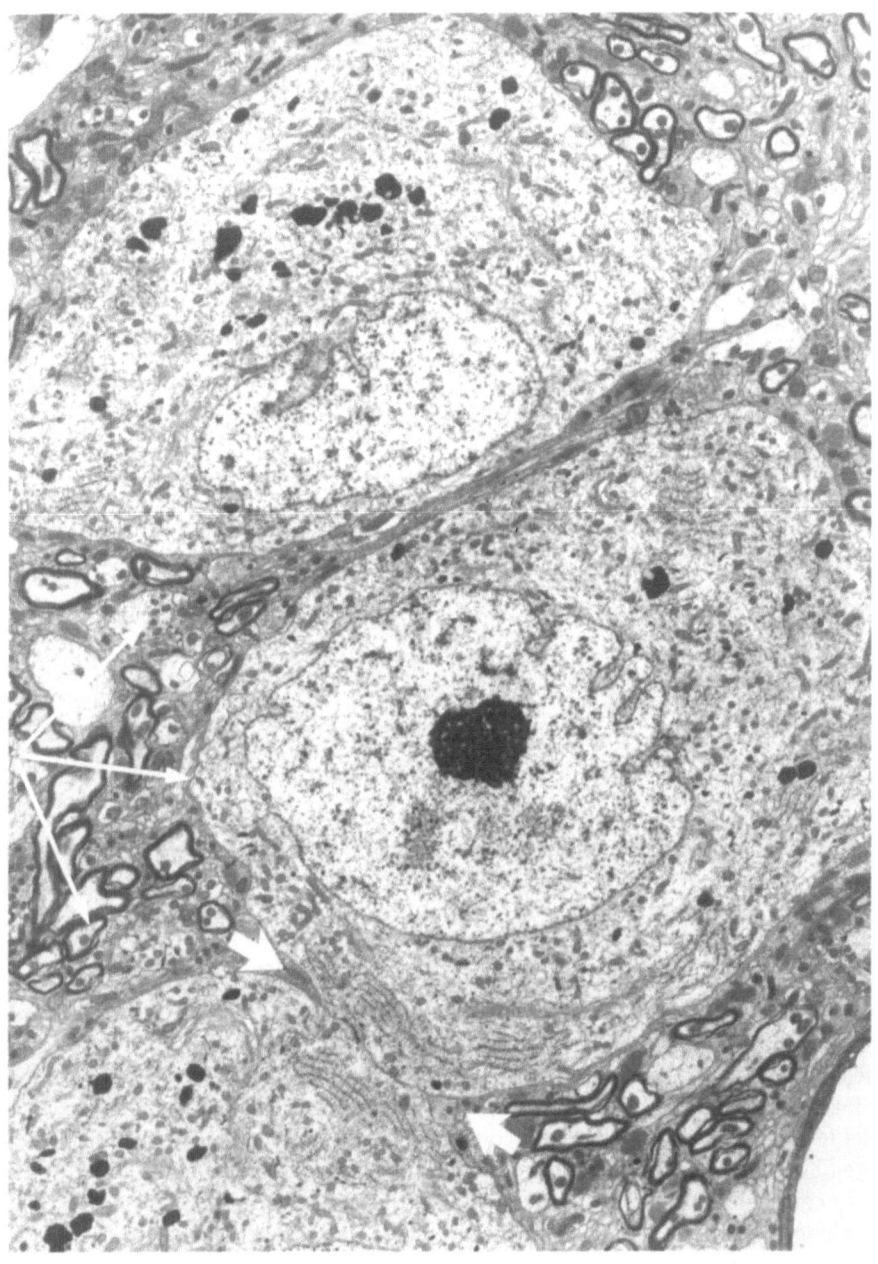

Fig. 10. A group of three neurons (*long white arrows*) in the cat inferior olivary complex. Two adjacent neurons in apposition (*short white arrows*). ×4700

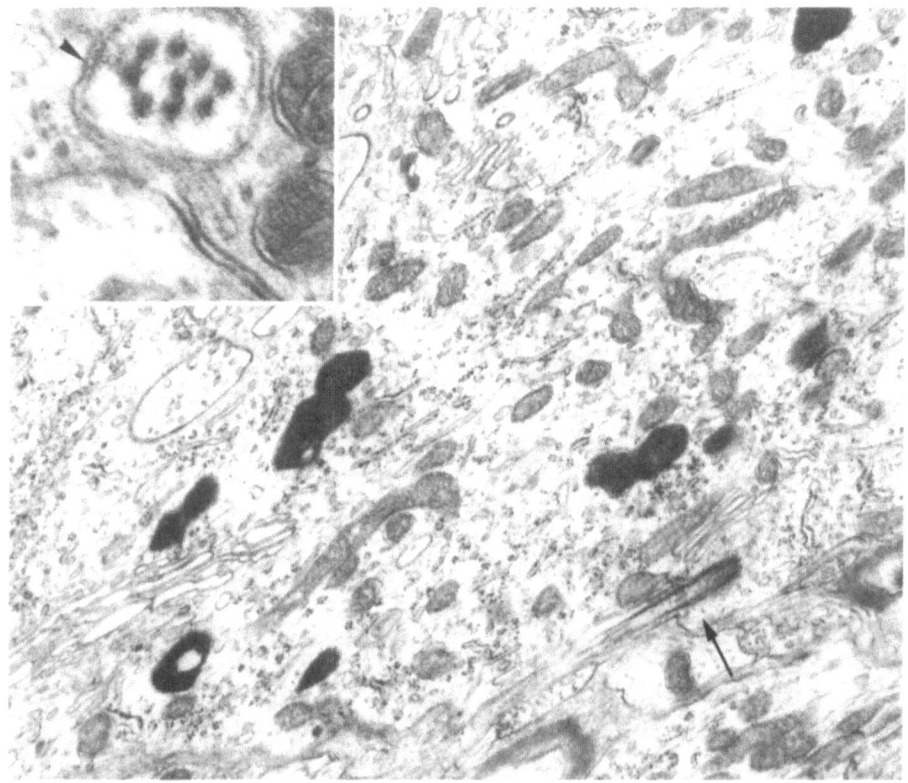

Fig. 11. A ciliated neuron with a longitudinally sectioned cilium (*arrow*). Ground squirrel. ×11,500. The inset illustrates a transverse sectioned cilium with an 8+1 microtubular pattern (*arrowhead*). ×41,400

on neuronal somata and proximal dendrites. The second type of axon terminal contains pleomorphic vesicles and forms mainly symmetric synaptic junctions with neuronal somata and somatic spines. Both types of axon terminals contain dense core vesicles in addition to other synaptic vesicles. These types of axon terminals occasionally form crest synapses with the postsynaptic membrane of somatic spines.

Large-sized reticular neurons are considered as a special type. These neurons are located between olivary neurons or at the boundaries of all subnuclei in the inferior olivary complex of the electron-microscopically examined species. Their perikaryon is of elongated shape and rich in organelles. The number of axosomatic synapses is higher in the larger reticular cells than in typical olivary neurons. Two types can be distinguished among the axosomatic terminals. The first type of axon terminal contains pleomorphic synaptic vesicles and form symmetric synaptic junctions. The second type of axosomatic terminal contains round synaptic vesicles and form asymmetric, intermediate, or symmetric synaptic junctions. Reticular neurons bear somatic spines with a wide variety of sizes and shapes. One or two synaptic terminals containing rounded or flattened vesicles with mainly symmetric synaptic junctions contact these spines.

24

3.2.2
Neuropil

The neuropil of the inferior olivary complex in every species and all examined animals is constituted by dendritic profiles, myelinated and unmyelinated axons, axon terminals or varicosities, and many glial cells and processes.

3.2.2.1
Dendrites

In olivary neurons, primary dendrites arise from the cell body. Many of the organelles of the perikarya extend into the aspinal segment of the proximal dendrite. The base of the proximal dendrite of olivary neurons, mainly of second or ball-like type, is in continuity with the initial axon segment. The initial axon segment is identified by fasciculation of microtubules and a dense membrane undercoating. The initial axon segment is usually thin (1 μm) and forms no synaptic contacts. Proximal dendrites contain the same constituents of cytoplasmic organelles as the perikaryon. The dendritic surfaces are mostly smooth with occasional stubby spines or dendritic appendages, which are contacted by synaptic terminals. As in the perikarya, few axon terminals with pleomorphic vesicles synapse on their plasmalemma. Astroglial processes cover most parts of these dendrites.

The secondary and tertiary dendrites (less than 2 μm–1 μm in diameter) are characterized by few cytoplasmic organelles including free ribosomes, smooth endoplasmic reticulum, mitochondria, and microtubules. Some dendritic profiles are filled with numerous mitochondria. Dense-core vesicles, multivesicular bodies, and dendritic lamellar bodies are present in the cytoplasm of dendrites. These lamellar bodies are found in the inferior olivary complex of the rat and cat, but in ground squirrels only cytoplasmic lamellar bodies in connection with cisterns of the granular endoplasmic reticulum and subsurface cisterns could be identified. Some dendrites sectioned in the longitudinal plane follow a curved path and exhibit swellings between thin intersegments. The olivary dendrites occasionally form direct dendro-dendritic appositions without any specializations, so-called casual appositions. The dendro-dendritic appositions can include puncta adhaerentia, either solitary in all species examined or in combination with a gap junction, which were only found in the mammalian inferior olivary complex.

In thin sections, some dendritic profiles exhibit a small number of spines such as stubby spines, branched or stalked spines with slender necks and bulbous heads. Dendrites in the inferior olivary complex in all animal examined are receptive sites for synaptic terminals or varicosities. Convergence and divergence in the inputs are evident. Dendritic profiles containing clear synaptic and dense core vesicles occasionally form synaptic contacts with synaptic terminals (Fig. 12).

3.2.2.2
Synapses

The synaptic population in all subnuclei of the inferior olivary complex of all animals examined is heterogeneous, containing two main classes of axon terminals or varicosities.

The first class contains predominately round synaptic vesicles (mean diameter 38.8±0.6 nm, SEM). This class includes terminals with various sizes (less than

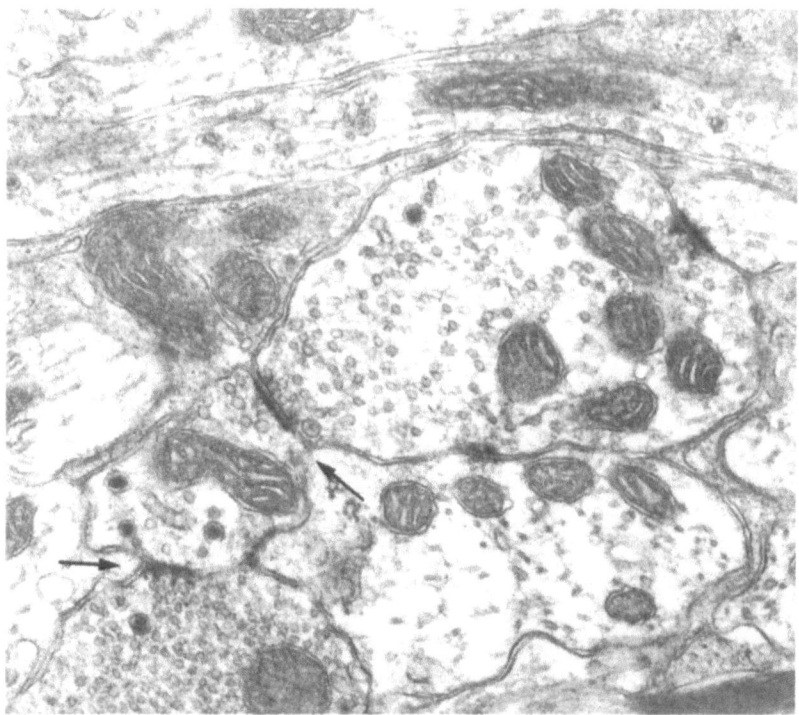

Fig. 12. The small dendrite having rounded vesicles and dense core vesicles (*arrows*) forms two symmetric synapses with synaptic terminals containing rounded or pleomorphic vesicles. Ground squirrel. ×11,500

1–4.5 μm in diameter). The small to medium-sized terminals form asymmetric synaptic junctions mainly on small and medium-sized dendrites and rarely on a perikaryon or on proximal dendrites. In addition to round synaptic vesicles, they contain populations of dense-core vesicles (diameter 60–80 nm). This type of axon terminal occasionally exhibits subsynaptic dense bodies beneath the postsynaptic membrane. Small to medium-sized terminals, with irregular round vesicles, form crest synaptic junctions with dendritic spines and slightly asymmetrical junctions with the dendritic shaft. We did not observe crest synaptic junctions in the inferior olivary complex of the pigeon. Finally, there are examples in which one large axon terminal containing round vesicles (38 nm in average diameter) forms several synaptic junctions of asymmetrical, intermediate, or slightly symmetrical type with one dendritic profile. However, the typical perforated synapses are not found.

The second class is composed of profiles containing pleomorphic synaptic vesicles, a mixture of round, oval, and slightly flattened vesicles, which range in size. These axon terminals often contain dense-core vesicles (diameter 60–80 nm). They mainly constitute symmetrical synaptic junctions with dendrites of all size, dendritic spines, and neuronal somata. This type of synaptic terminal occasionally makes up symmetrical contacts with one dendritic profile and at the same time introduces asymmetrical synaptic junctions with a second dendrite exhibiting subsynaptic dense bodies. Close to the asymmetrical synaptic junction, puncta adhaerentia in combination with omega-shaped depression are present. These examples deviate from the rule that synaptic terminals with round vesicles form asymmetrical synaptic junctions and those with pleomorphic vesicles form symmetrical junctions. Finally, the synaptic terminals containing pleomorphic and dense-core vesicles constitute crest synaptic junctions with somatic or dendritic spines.

In the neuropil of the inferior olivary complex, some axon terminals and boutons en passant contain a large number of dense-core vesicles but do not form specializations with other elements they are in apposition with.

3.2.2.3
Special Synaptic Arrangement

In the neuropil of the inferior olivary complex there is an important synaptic arrangement, called glomerulus, that is widely present in all subnuclei of this complex in rats, ground squirrels, and cats. In these glomeruli or complex synaptic clusters, dendritic profiles form a central core for the assembly of several axon terminals or varicosities, which are enveloped by astrocytic lamellae (Fig. 13). The central core of this complex synaptic arrangement is composed of 5–12 dendritic elements with a mean diameter of 0.605±0.06 µm (SEM). These dendritic elements presumably stream from different dendrites. The Golgi-impregnated dendritic profile with dendritic appendages that are gold-toned can be easily traced in serial thin sections. These dendritic appendages are found in apposition with non-impregnated profiles in different complex synaptic clusters (Fig. 14). In some sections, the connections between the parent dendritic profile and its appendages are visible. These dendritic appendages emerge from tertiary dendrites near the periphery of the olivary complex synaptic field. On the appendages, dendro-dendritic appositions feature by the gap junction. An omega-shaped invagination and puncta adhaerentia are present close to the gap junction (Fig. 15). Dendritic profiles in the glomerular central core can be infrequently originated from primary dendrites and the perikaryon.

Two main classes of axon terminals and boutons en passant can be distinguished in the periphery of olivary complex synaptic fields. One class contains predominantly rounded synaptic vesicles and the second class of axon terminals has pleomorphic vesicles. The pleomorphic vesicles consist of a mixture of rounded, oval, and some flattened vesicles. Axon terminals containing only flattened vesicles are not present. In addition to round and pleomorphic vesicles, these terminals often contain dense-core vesicles. Axon terminals with predominantly rounded vesicles form asymmetrical synaptic junctions mainly on small to medium-sized dendrites and dendritic spines, while the second class of axon terminals forms mainly symmetrical synaptic junctions on the neuronal somata and dendritic profiles. In many cases, analysis of sections of both classes of axon terminals revealed that they form either symmetrical or asymmetrical synaptic junctions without any consistent association between the shape of the vesicles and the type of synaptic junction. For example, the axon terminal with rounded synaptic vesicles, which is sectioned longitudinally, forms symmetrical synaptic junctions with dendritic profiles in dendro-dendritic appositions, including a gap junction between them (Fig. 16). In some cases, two axon terminals containing pleomorphic vesicles form a crest synaptic junction at the neck of a dendritic spine of the glomerular central core and make up a symmetric synaptic junction and puncta adhaerentia, with the tertiary dendrite giving rise to this dendritic spine (Fig. 17). The axon terminals in the glomerular periphery range in size from small (diameter

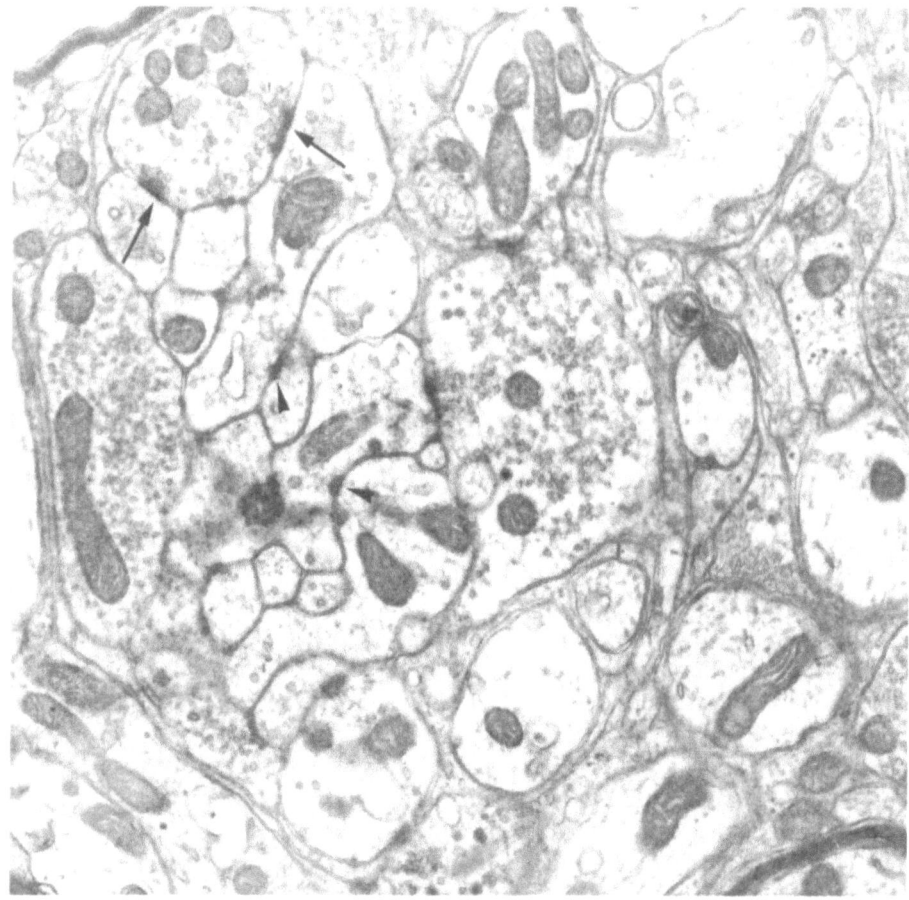

Fig. 13. The complex synaptic junction (glomerulus) is capsulated by astrocytic lamellae. The core is composed of dendritic appendages and some of them exhibit puncta adhaerentia (*arrowheads*). The axon terminals with pleomorphic vesicles form symmetrical synapses (*arrows*). Ground squirrel. ×15,250

0.9 μm) to medium-sized (diameter 2.1 μm). Small axon terminals usually form short synaptic junctions per dendrite, but large terminals can make synaptic contacts with three or more different dendritic profiles.

The complex synaptic clusters, with a central core of dendritic profiles that are connected by puncta adhaerentia, are present in the pigeon's inferior olivary complex, but until now dendro-dendritic gap junctions have not been observed.

3.2.2.4
Intramembranous Structure of the Synaptic Contact Zone and Dendro-Dendritic Junctions

In freeze-etching replicas, the synaptic contact zone is unequivocally identified when synaptic vesicles are exposed simultaneously with part of the presynaptic membrane. Large intramembranous particles, numerous pits, and various numbers of smooth bumps or protuberances at its slightly convex aspect (Fig. 18) characterize the E-face

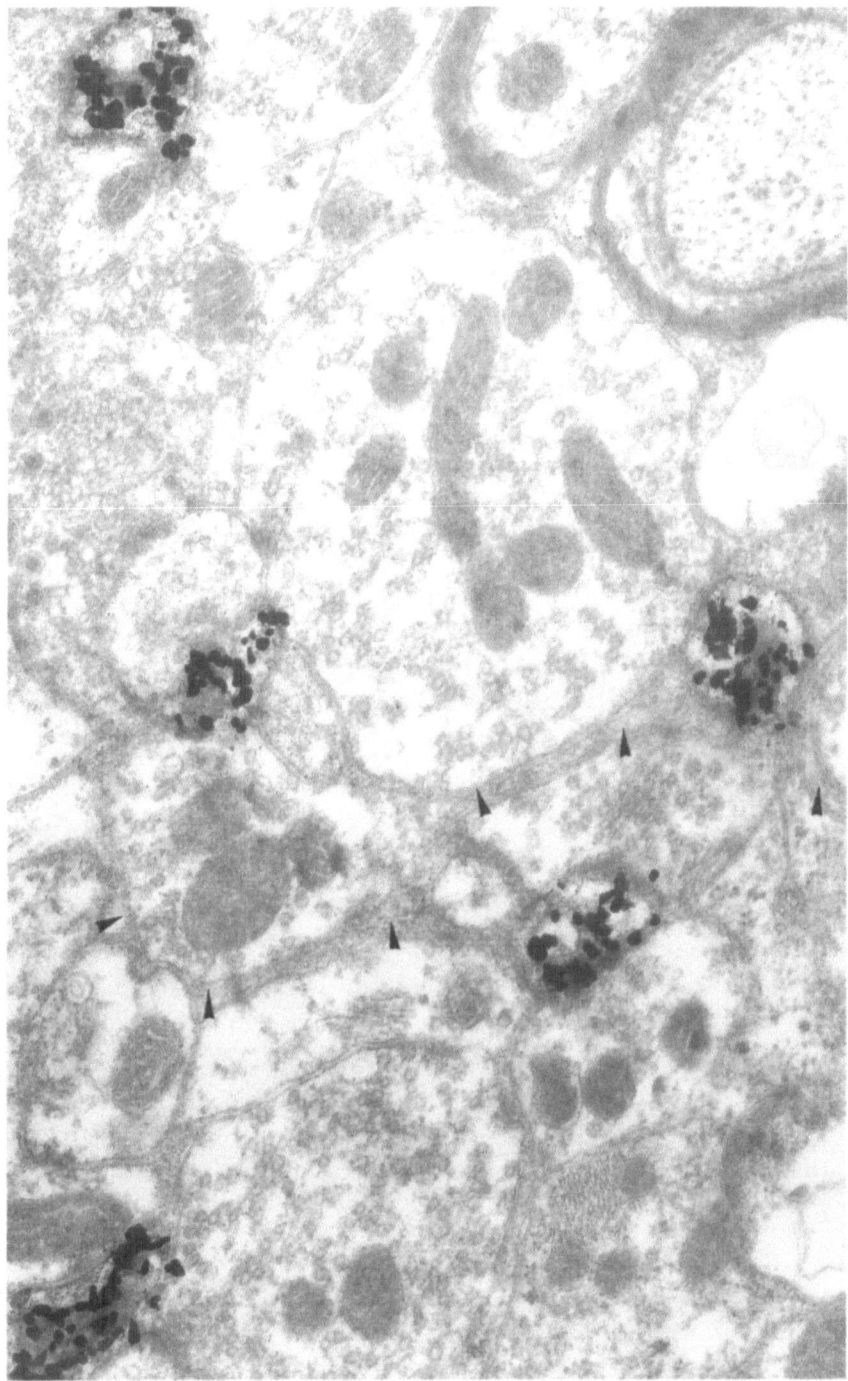

Fig. 14. The Golgi-impregnated dendritic appendages located in different complex synaptic junctions are separated by a marked astrocytic lamella (*arrowheads*). Ground squirrel. ×34,500

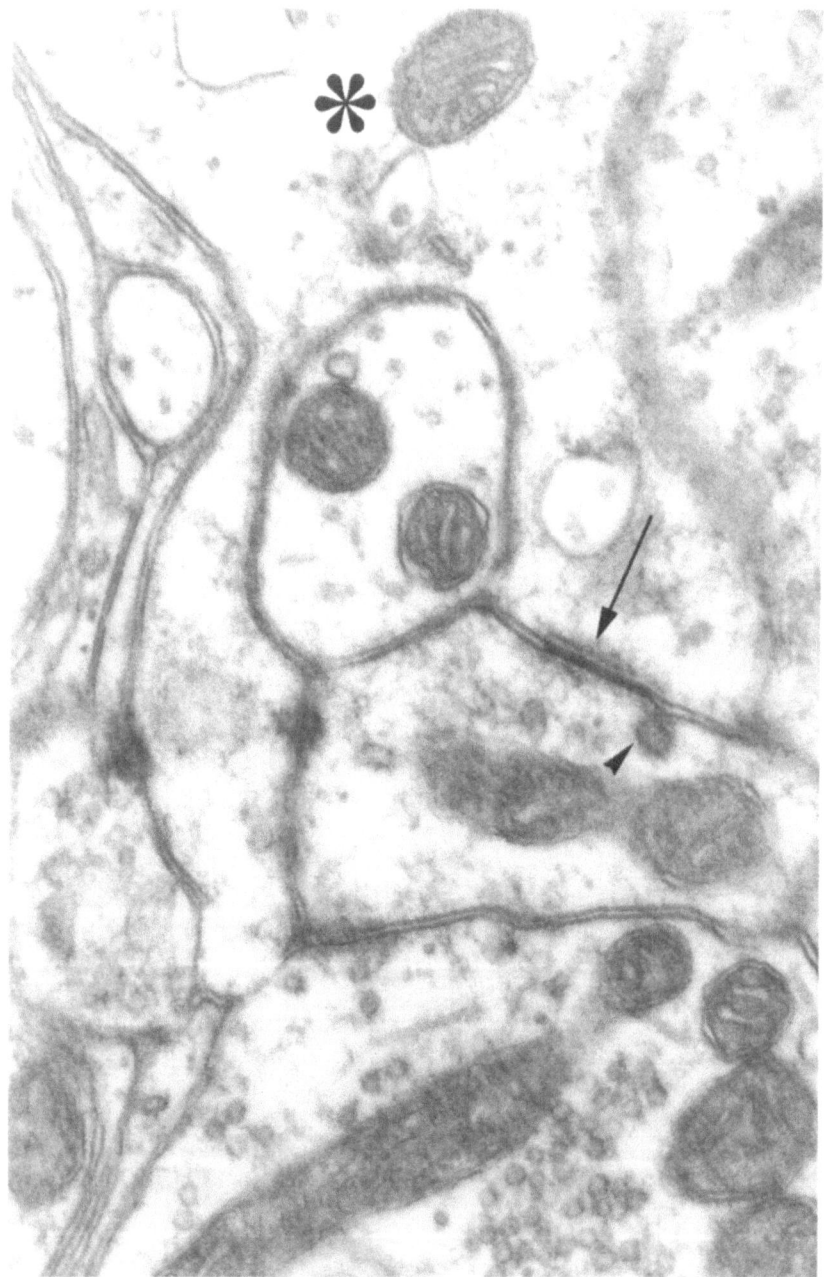

Fig. 15. Both dendritic protrusions originate from the parent dendritic profile (*asterisk*). A gap junction (*arrow*) links these appendages. The omega-like structure (*arrowhead*) appears as a crater. ×62,100

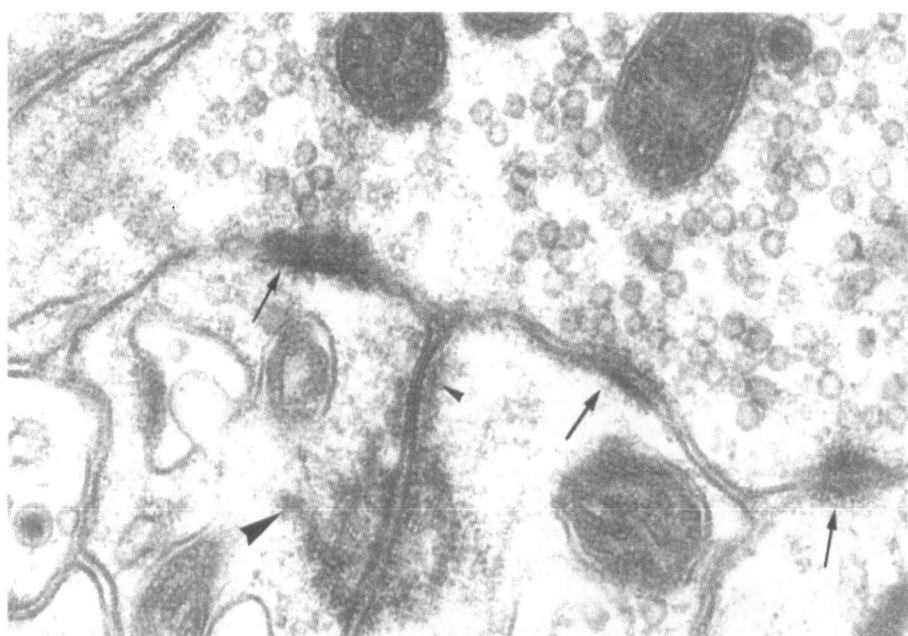

Fig. 16. The dendritic profiles are linked by a gap junction (*arrowhead*) and puncta adhaerentia (*large arrowhead*). The axon terminal containing rounded vesicles forms symmetric synaptic contacts with the dendritic appendages (*arrows*). ×110,000

of the presynaptic membrane. In this case, the fracture process shifted from the presynaptic membrane to the postsynaptic membrane and simultaneously exposed the presynaptic E-face specializations and the postsynaptic P-face of the dendritic shaft membrane.

At the synaptic contact zone, the presynaptic P-face is slightly concave and appears round or oval-shaped. The intramembranous particles are more numerous and larger than the adjacent non-junctional membrane. These intramembranous particles tend to vary in size and distribution among the presynaptic modulations, called dimples (Fig. 19). In most synaptic contact zones, the presynaptic membrane usually has the appearance of presynaptic modulations at both fracture faces, but in other cases appears to be undulated. Some presynaptic membranes have large intramembranous particles on the P-face and their contours appear to be depressed at the synaptic contact zone. Other presynaptic membranes are relatively flat with uniformly distributed intramembranous particles on the P-face. These differences in intramembranous structure of the presynaptic membrane are probably the result of differences in the rate of transmitter release and vesicle exocytosis during the period of fixation, or depend on the types of axons involved.

Two types of internal organization characterize the E-face of the postsynaptic membrane. Firstly, at the synaptic contact zone, the E-face of the postsynaptic membrane can be free of intramembranous particles in apposition to the presynaptic membrane specializations. According to generally recognized criteria (Landis and Reese 1974), this is the freeze-etching equivalent of the symmetric or inhibitory synaptic junction. The E-face of the postsynaptic membrane is positively identified

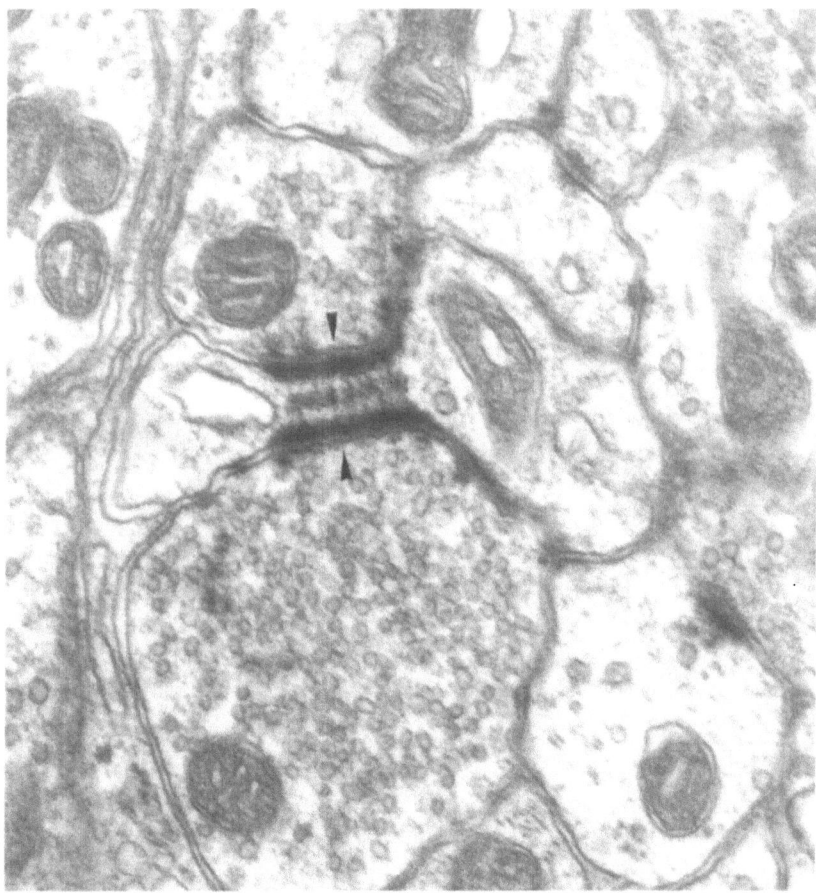

Fig. 17. Electron micrograph of the olivary glomerulus. Two axon terminals containing pleomorphic vesicles form the crest synaptic junction (*arrowheads*) of the neck of the dendritic protrusion. Ground squirrel. ×59,000

when parts of the postsynaptic membrane are simultaneously exposed with presynaptic modulations on the presynaptic P-face.

Secondly, at the synaptic contact zone, the E-face of the fractured dendritic membrane has aggregates of intramembranous particles, which vary in size and probably correspond to the asymmetrical synaptic or excitatory junctions. The postsynaptic membrane is positively identified when the fracture process shifts from the postsynaptic E-face to the presynaptic P-face and expresses simultaneously presynaptic P-face dimples among the intramembranous particles and postsynaptic E-face aggregates on the adjacent synaptic contact zone (Fig. 20). These small or sometimes larger aggregates show uniformly distributed intramembranous particles in a flat, concave, or convex macular pattern. In the pigeon inferior olivary complex, horseshoe-like aggregates may also, on rare occasions, occur on the postsynaptic E-face (Fig. 21). The postsynaptic E-face, however, is rarely found. Both faces of the presynaptic membrane are predominately exposed.

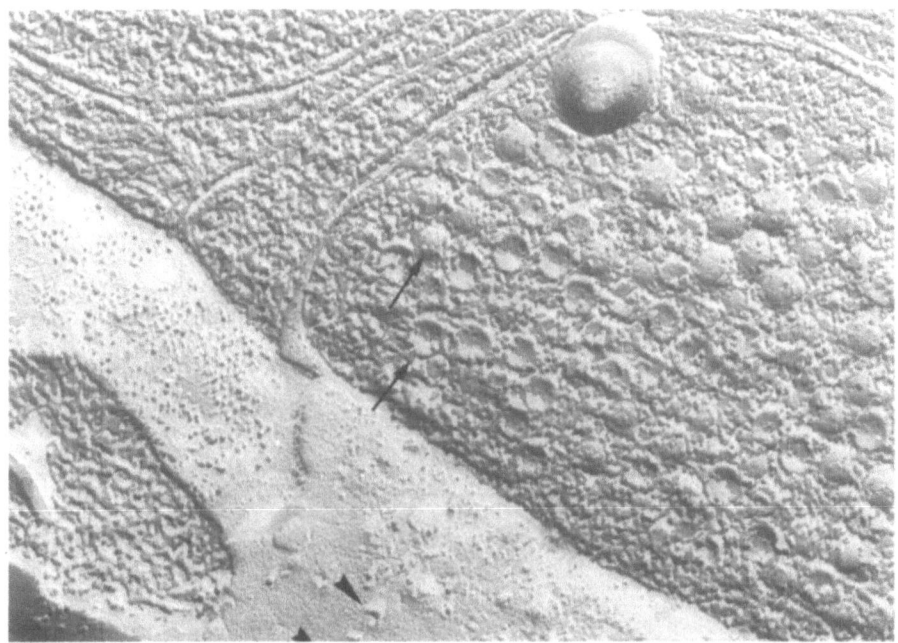

Fig. 18. The presynaptic E-face at the synaptic contact zone is identified by the presence of large intramembranous particles, protuberances (*arrowheads*), and cross-fractured synaptic vesicles in the cytoplasm (*arrows*). Rat. ×78,200

The average size of intramembranous particles at the E-face of the postsynaptic membrane is 10.7±0.44 nm (SEM). The average size of intramembranous particles on the E-face of the presynaptic membrane is 10.5±0.6 nm (SEM) and 11.8±0.57 nm (SEM) on the presynaptic P-face. The comparisons of the average sizes of the intramembranous particles in postsynaptic E-face aggregates, presynaptic E- and P-face particles are not significantly different ($p>0.10$). On the contrary, the particle packing density in the postsynaptic aggregates is significantly different from that on the E- and P-face of the presynaptic membrane ($p<0.001$)

Both faces of the presynaptic membrane contain intramembranous particles and membrane modulations in apposition to the specialized or unspecialized E-face postsynaptic membranes, which may correspond to symmetrical and asymmetrical synaptic junctions in the inferior olivary complex.

In freeze-etch replicas, puncta adhaerentia are characterized by intramembranous particles on both faces of the dendritic membranes (Fig. 22). During the freeze-etch procedure the lipid bilayer of membranes is cleaved along the hydrophobic inferior. The E-face exposes the extracellular half of the lipid bilayer while the P-face exposes the cytoplasmic half within the domain of the puncta adhaerentia. Dendro-dendritic puncta adhaerentia are unequivocally identified because the P-face has a higher density of intramembranous particles than the E-face.

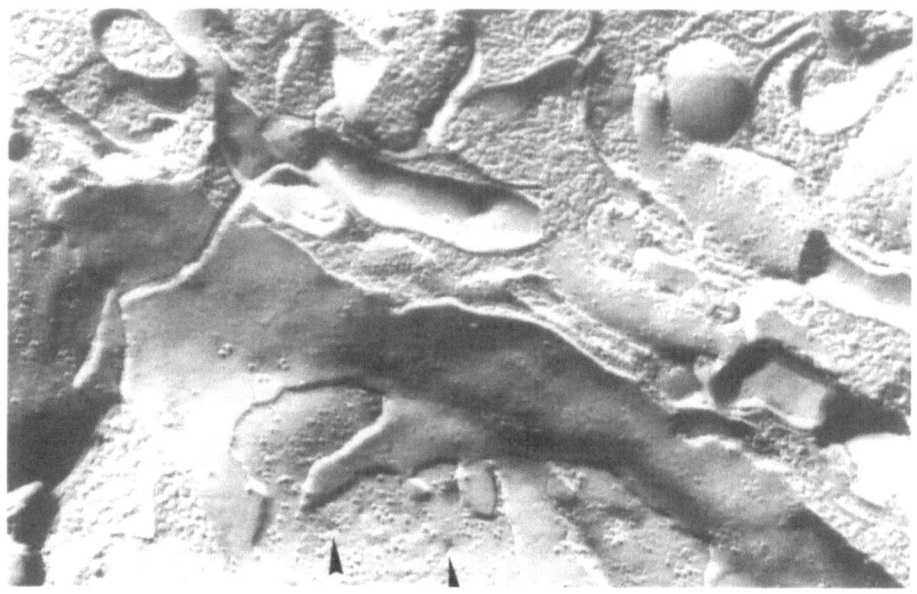

Fig. 19. Specializations on the P-face of the presynaptic membrane. A gentle concavity, large intra-membranous particles, and dimples (*arrowheads*) characterize the synaptic contact zone. Rat. ×57,600

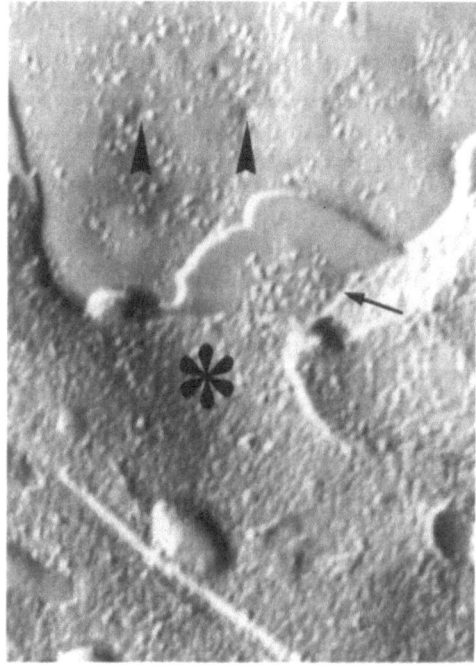

Fig. 20. The fracture plane splitting the postsynaptic membrane with part of the cytoplasm (*asterisk*) and the E-face aggregate of the intramembranous perticles (*arrow*) is co-extensive with the presynaptic P-face specializations (*dimples, arrowheads*) on the adjacent synaptic contact zone. Cat x 54,600

Fig. 21. A horseshoe-shaped aggregate of the intramembranous particles (*arrows*) characterizes the E-face of the postsynaptic dendritic spine membrane. Pigeon. ×84,000

The P-face is rich in intramembranous particles, which are larger, more prominent, and randomly distributed. The E-face also possesses randomly distributed intramembranous particles, but they are few.

In freeze-etching replicas, dendro-dendritic gap junctions are easily recognized when the fracture process shifts from the first to the second dendritic membrane, and the E- and P-face of the adjacent dendrite membranes are exposed side by side (Fig. 23). In this example, the gap junction is unequivocally identified because the closely packed P-face particle aggregate is exposed simultaneously with arrays of pits at the adjacent dendritic E-face, across a narrowing of intercellular space. This gap junction is large and its particles are loosely arranged with small islands of smooth membrane. At the P-face, gap junction particles are of uniform dimensions, at about 10 nm, and the particle packing density is 4430 ± 127 particles per μm^2. At the synaptic contact zone, the greatest particle packing density in the postsynaptic E-face aggregates is significantly different to that in the gap junctions ($t=9.05$, $p<0.001$). Figure 23 illustrates intramembranous particles in the E-face depression situated at a short distance of the gap junctional domain. This is interpreted as an omega-shaped invagination. The presynaptic E-face is exposed simultaneously with the gap junctional domain and contains intramembranous particles and smooth bumps.

In some freeze-etching replicas two dendritic protrusions are cut in continuity with their parent dendritic stem (Fig. 24). The uniform dimensions of their subunits characterize the aggregation of gap junction particles on the P-face of one of these dendritic protrusions. This gap junction aggregate is small and its particles are tightly packed in a hexagonal pattern and are delimited from the surrounding area. This dendritic profile forms a synaptic contact with an axonal terminal, which is presynaptic to the dendritic protrusions of the adjacent dendritic stem. The synaptic contact

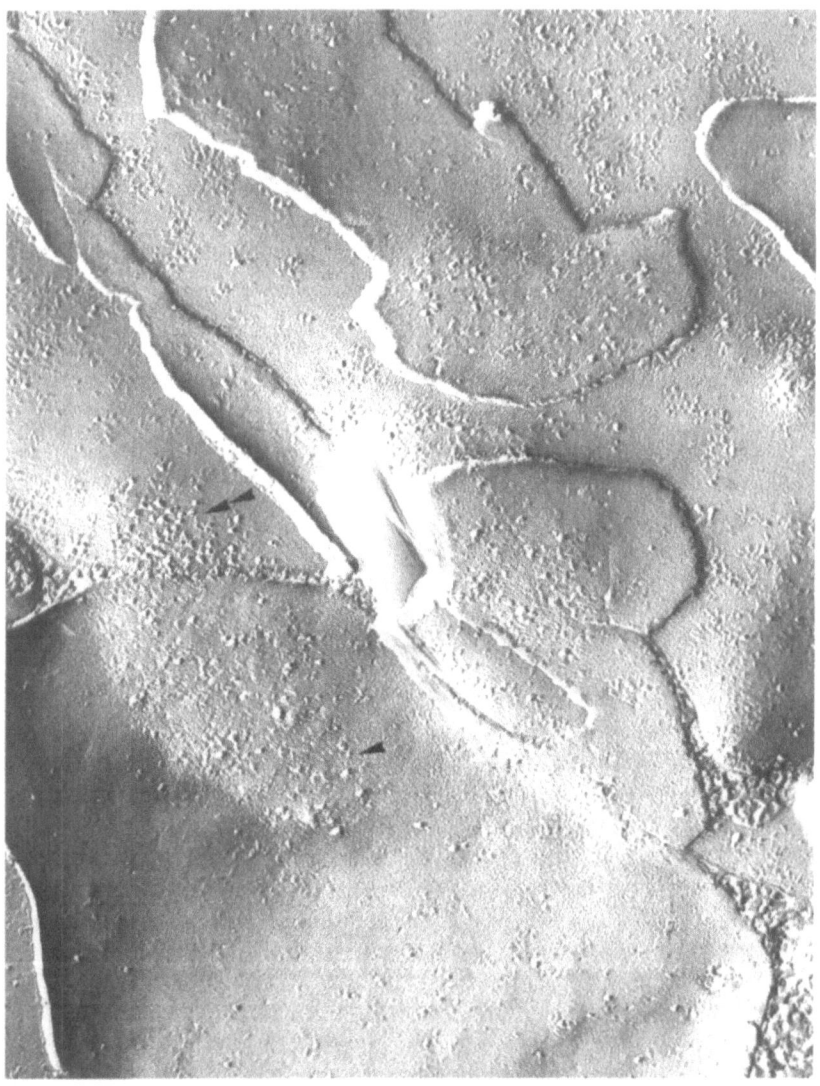

Fig. 22. Dendro-dendritic puncta adhaerentia. The dendritic membranes, the small area of the P-face (*double arrowheads*), and the large area of the E-face (*arrowhead*) are exposed. Pigeon. ×46,800

zone at the presynaptic P-face is exposed simultaneously with the synaptic cluster of vesicles in the cross-fractured axonal terminal and with parts of postsynaptic E-face. At the synaptic contact zone the presynaptic P-face is characterized by intramembranous particles among the dimples, but the postsynaptic membrane on the E-face is unspecialized. This is probably the freeze-etching equivalent of the symmetric synaptic junction. At the same time, the dendrite shaft makes a synaptic contact, while intramembranous particles, pits, and protuberances characterize the presynaptic E-face.

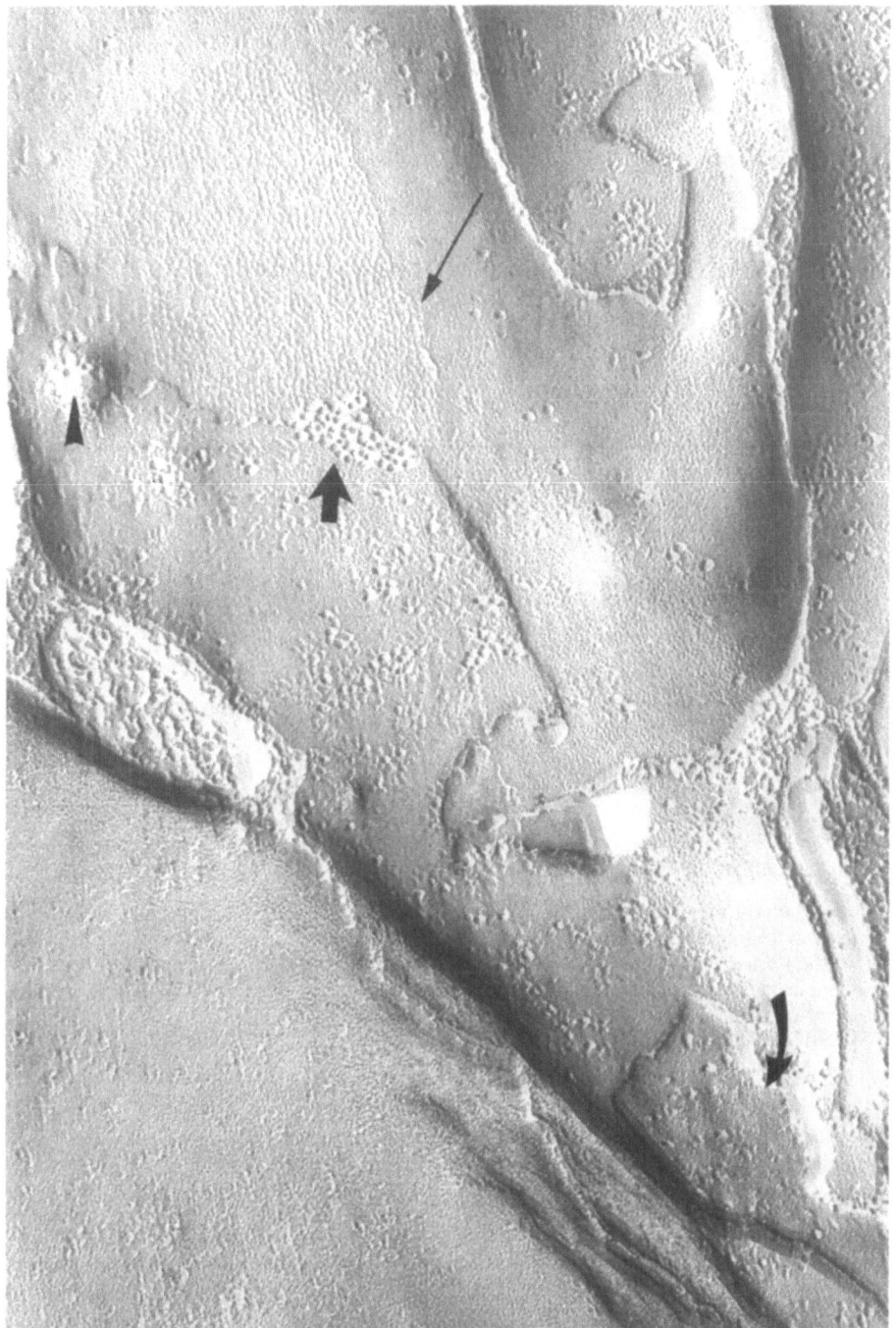

Fig. 23. Dendro-dendritic gap junction. The plan of the fracture shows the intramembranous particles on the P-face (*short arrow*) and the complementary pits on the E-face (*long arrow*). The coated pit appears as a crater (*arrowhead*) on the P-face near the gap junction. The presynaptic E-face (*curved arrow*) is exposed simultaneously with a gap junctional domain and contains intramembranous particles and smooth bumps. Ground squirrel. ×153,000

Fig. 24. Uniform dimensions of subunits characterize the gap junction aggregate particles (*short arrows*) on the P-face of a dendritic protrusion. The E-face of this dendritic protrusion is postsynaptic to the presynaptic P-face of the cross-section axon terminal (*arrowheads*). The postsynaptic membrane on the E-face is unspecialized. The dendritic protrusion is in continuity with a parent dendritic stem (*long arrow*). Ground squirrel. ×48300

3.2.2.5
The Neuroglial Cells

Neurons in the inferior olivary complex in all examined animals tend to be arranged in groups and are embedded in a large number of myelinated axons and glial cells. In Golgi preparations, the surrounding white matter is populated by fibrous astrocytes. This type can be seen in close apposition to the myelinated bundles in combination with protoplasmic astrocytes and oligodendrocytes. True protoplasmic astrocytes have a lager cell body and their radial processes contain small protrusions. They can be in a perineuronal or a perivascular position. The staining patterns obtained by GFAP of these astrocytes containing gliofilaments are comparable with those obtained by the Golgi method (Fig. 25a). Glial processes can be followed a long distance away from the cell body.

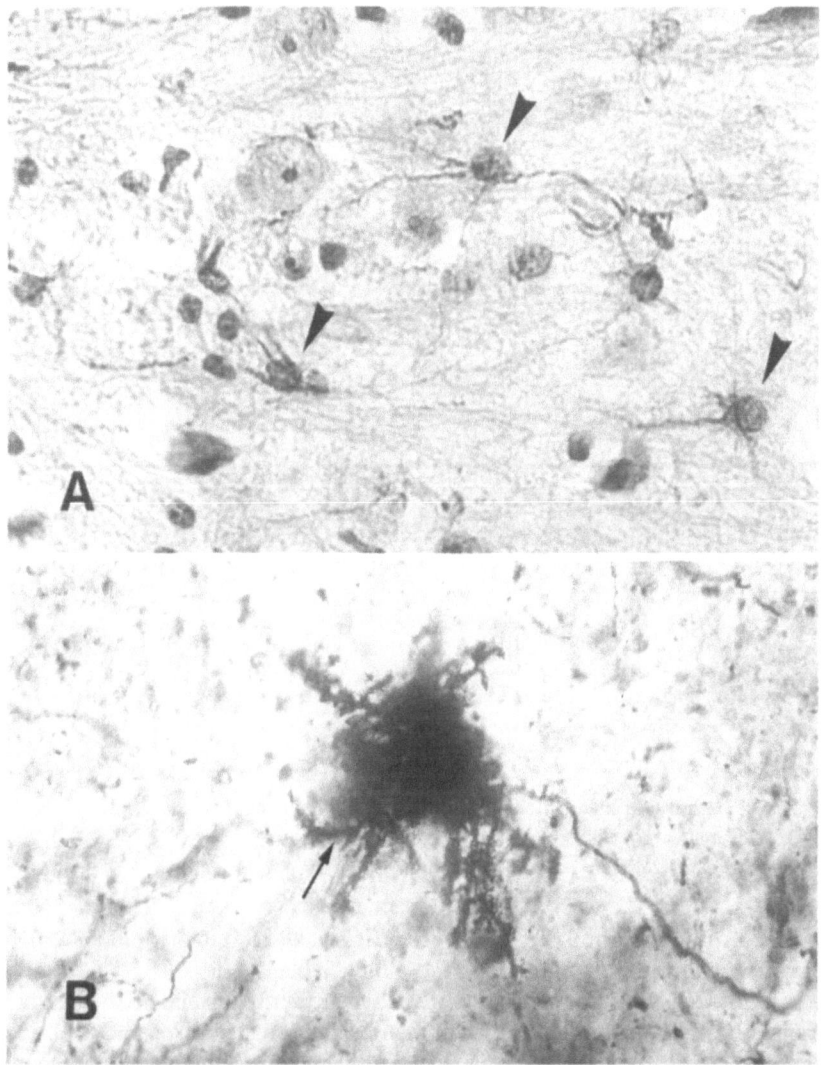

Fig. 25. a GFAP-positive astrocytes (*arrowheads*) in the ground squirrel. Contra-staining with hematoxylin, ×397. **b** Golgi-impregnated velate astrocyte (*arrow*) in the ground squirrel. ×712

The velate astrocytes with divided processes and velamentous projections are seen in the neuropil of all examined animals, including humans. They are the predominate astrocyte type in the inferior olivary complex of the ground squirrel (Fig. 25b). In ultrathin sections, fibrous astrocytes contain numerous bundles of gliofilaments running parallel into its main branch. Protoplasmic astrocytes contain only a few gliofilaments extending into the main processes (Fig. 26). Processes of these two astrocyte types end against the blood capillaries with an end-foot. The velate type of astrocyte contains sheet-like processes or lamellae. These lamellae start directly from the cell body or the larger processes and are arranged as single or multilamellar formations

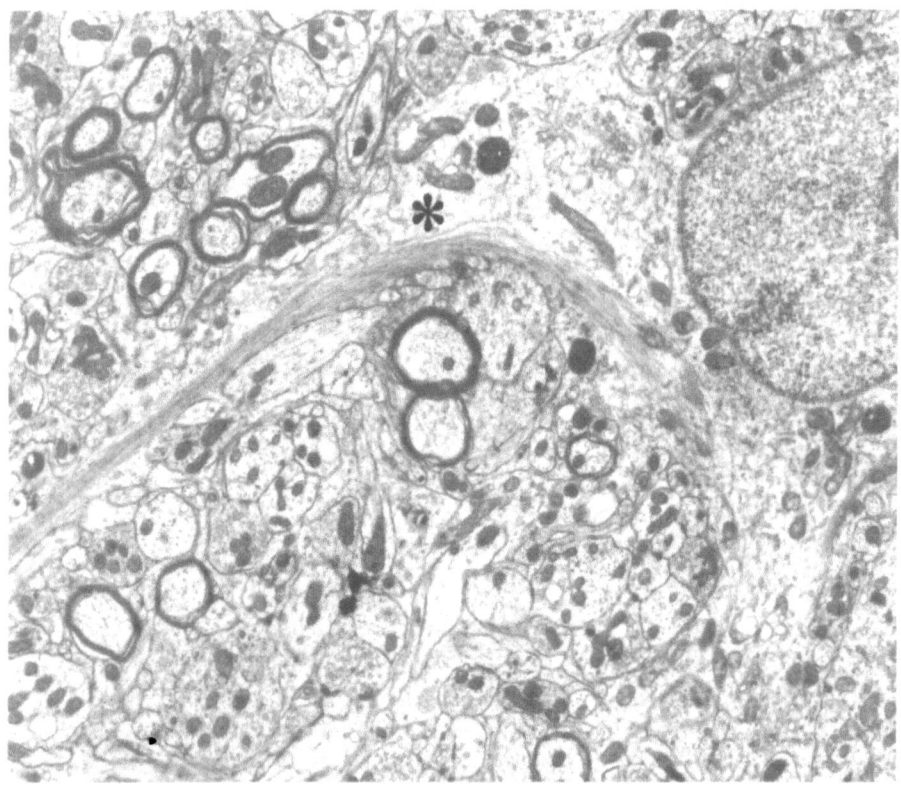

Fig. 26. The protoplasmic astrocyte (*asterisk*) with a few gliofilaments extending into the main processes. Ground squirrel. ×9,660

around elements of the neuropil (Fig. 27). Astrocytic–astrocytic appositions show two types of specializations, i.e., gap junctions and puncta adhaerentia.

In freeze-etch replicas, astrocytic processes are easily distinguished on the basis of their intramembranous ultrastructure. Their membranes are unequivocally identified by the presence of "assemblies" on the P-face and by complementary orthogonal arrays of pits on the E-face. These assemblies are composed of 5–15 or more intramembranous particles in orthogonal arrays that form square or rectangular aggregates on the P-face (Fig. 28a). In all animals which have been examined (pigeon, ground squirrel, rat, and cat) these arrays of intramembranous particles have about 41967 nm center-to-center periodicity. The background intramembranous particles measure 8–11 nm in diameter. The cross-fractured cytoplasm of the larger astrocytic processes contains bundles of gliofilaments (Fig. 28a). When the fracture process simultaneously exposes both the P- and E-faces of the astrocytic membranes, the aggregates of gap junction intramembranous particles on the P-face and complementary pits on the E-face are easily recognized (Fig. 28b). Intramembranous particles of gap junctions, the so-called connexons, are quite uniform (8–9 nm in diameter) and arranged in a hexagonal manner on the astrocytic P-face. The complementary pits are grouped in crystalline patterns and are associated with the astrocytic E-face. The membrane of the astrocytic lamellae in the multilamellar formation exhibits sparse

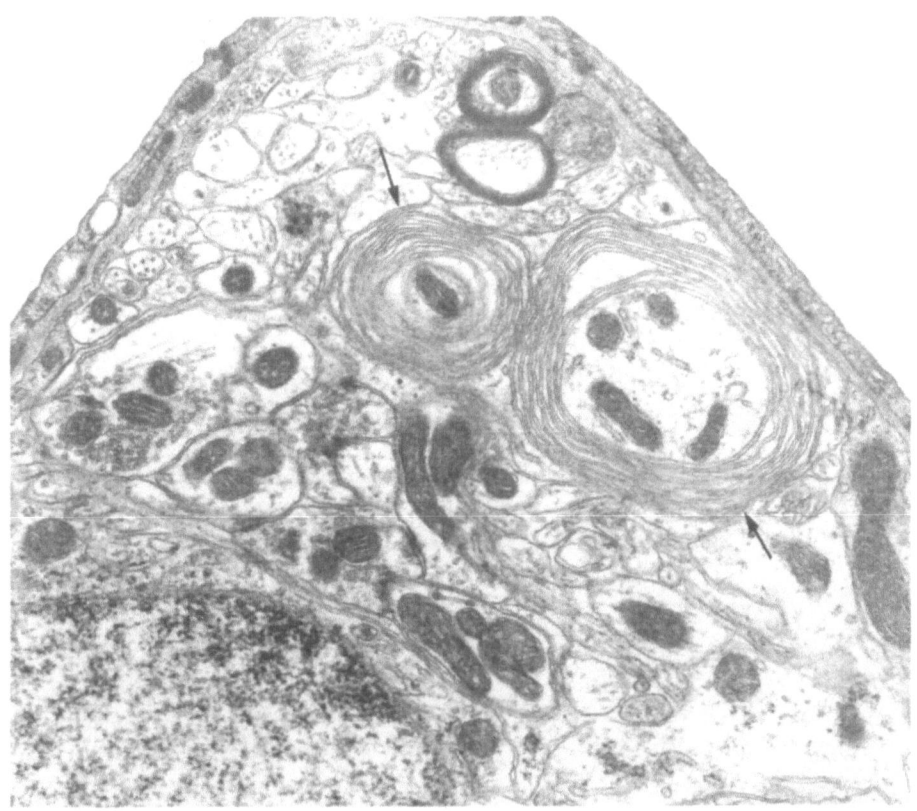

Fig. 27. Astrocytic lamellae (*arrows*) of the velate type. Ground squirrel. ×20,700

assemblies on the P-face (Fig. 29) and complementary pits on the E-face. The perivascular astrocytic membranes of the end-feet, however, contain a number of assemblies on the P-face (Fig. 30) and have complementary pits on the E-face.

Oligodendrocytes are small cells with round to oval or polygonal cell bodies and a few short processes. These cells have a round to oval nucleus in an eccentric position containing clamped peripheral chromatin and dense cytoplasm. Microtubules are the particular feature that distinguishes these glial cells from astrocytes. The cytoplasm is abundant with microtubules having a diameter of 24–25 nm that extend into processes and their outer or inner tongue. Oligodendrocytes can have a perineuronal position, a perineuronal-perifascicular (Fig. 31), or an interfascicular position. Oligodendrocytes and their processes form astrocytic–oligodendrocytic gap junctions and oligodendrocytic–oligodendrocytic tight junctions.

In freeze-etch replicas, the oligodendrocytic membrane of the cell body and its processes possess a heterogeneous population of intramembranous particles at both faces. The intramembranous particles are randomly distributed and are preferentially associated with the P-face. These particles tend to range in form and can be classified as globular or elongated (Fig. 32). They are numerous in portions near the compact myelin membrane. Some elongated particles (about 20–25 nm long) are composed of small portions and are arranged in linear strands. They are located at both faces of the

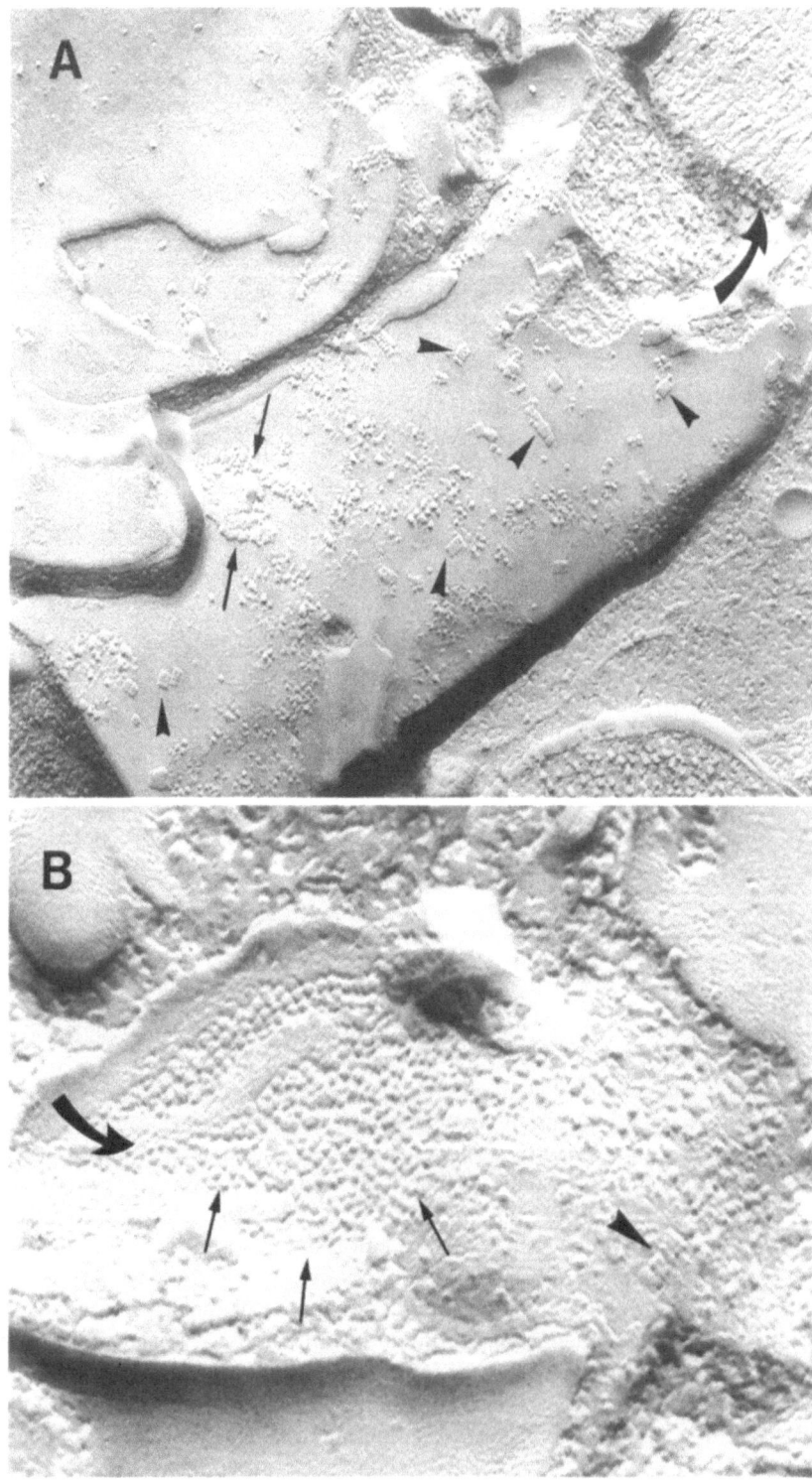

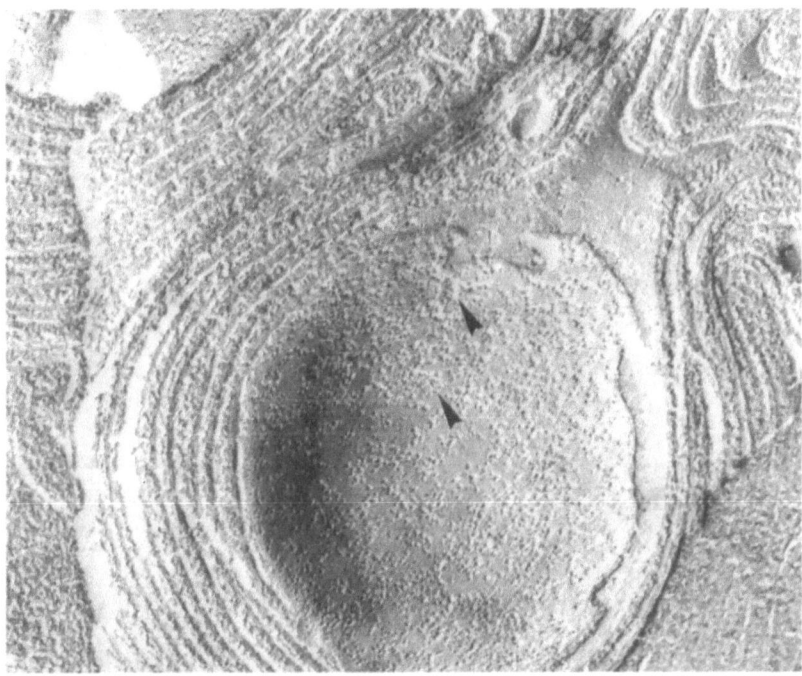

Fig. 29. Assemblies (*arrowheads*) on astrocytic lamella P-face within multilamellar capsule. Ground squirrel. ×34,400

oligodendrocytic membrane. The P-face of the oligodendrocytic membrane exhibits small dimples or larger invaginations which contain intramembranous particles. These invaginations probably correspond to the coated pits. Round and elongated pits are distributed at random among the intramembranous particles at both faces.

Oligodendrocyte-to-astrocyte processes (Fig. 32, inset), being appositions between cell bodies, between processes, or cell body and processes, include gap junction particles or connexons with an average size of 7.69±0.27 μm (SEM). These connexons are hexagonally packed and are associated with the P-face (Fig. 32). Pits on the E-face having a crystalline pattern (Fig. 33) complement them. The E-face contains plasmalemmal deformations as smooth bumps surrounded by intramembranous particles, which may be the result of the dimples at the P-face.

Oligodendrocyte-to-oligodendrocyte appositions include tight junctions. These linear specializations are located between oligodendrocytic cell bodies, or between

Fig. 28. a Cross-sectioned astrocytic processes containing bundles of gliofilaments (*curved arrow*). The assemblies on the P-face with a part of the complementary E-face pits (*arrowheads*) are distinct from gap junctional specializations (arrows). Cat. ×41,000. b Gap junction intramembranous particles are arranged in a hexagonal manner on the astrocytic P-face (*arrows*) with complementary pits on the astrocytic E-face (*curved arrow*). Assemblies (*arrowhead*) on the astrocytic P-face in the vicinity of the gap junction. Pigeon. ×118,000

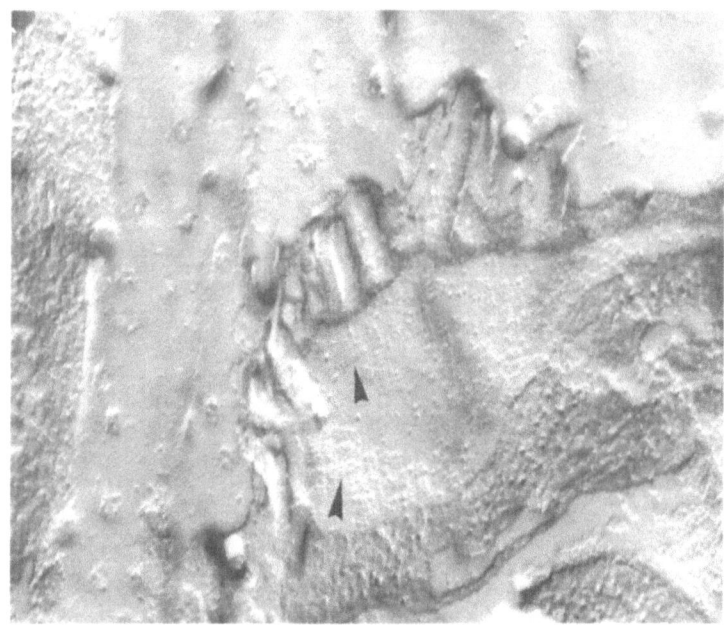

Fig. 30. The perivascular astrocytic membrane contains many assemblies on the P-face (*arrowheads*). Ground squirrel. ×34,400

cell body and process, as well as between the outer tongue process turn of the myelin sheath membrane. The P-face of the tight junctions is characterized by parallel ridges or strands composed of single and anastomosing intramembranous particles (Fig. 33). These particles measure about 8–10 nm in diameter. The E-face reveals grooves containing irregularly arranged individual intramembranous particles. These grooves appear to be complementary to the P-face ridges. The number of adjacent ridges or grooves on both faces and the distance between them are variable. We have not observed gap junctions at the oligodendrocytic membrane in the tight junction zone.

Oligodendrocytic-to-axon appositions in the paranodal region include bands of diagonal arrows of particles and parallel particle-free strands on the P-face of the oligodendrocytic membrane of paranodal pockets.

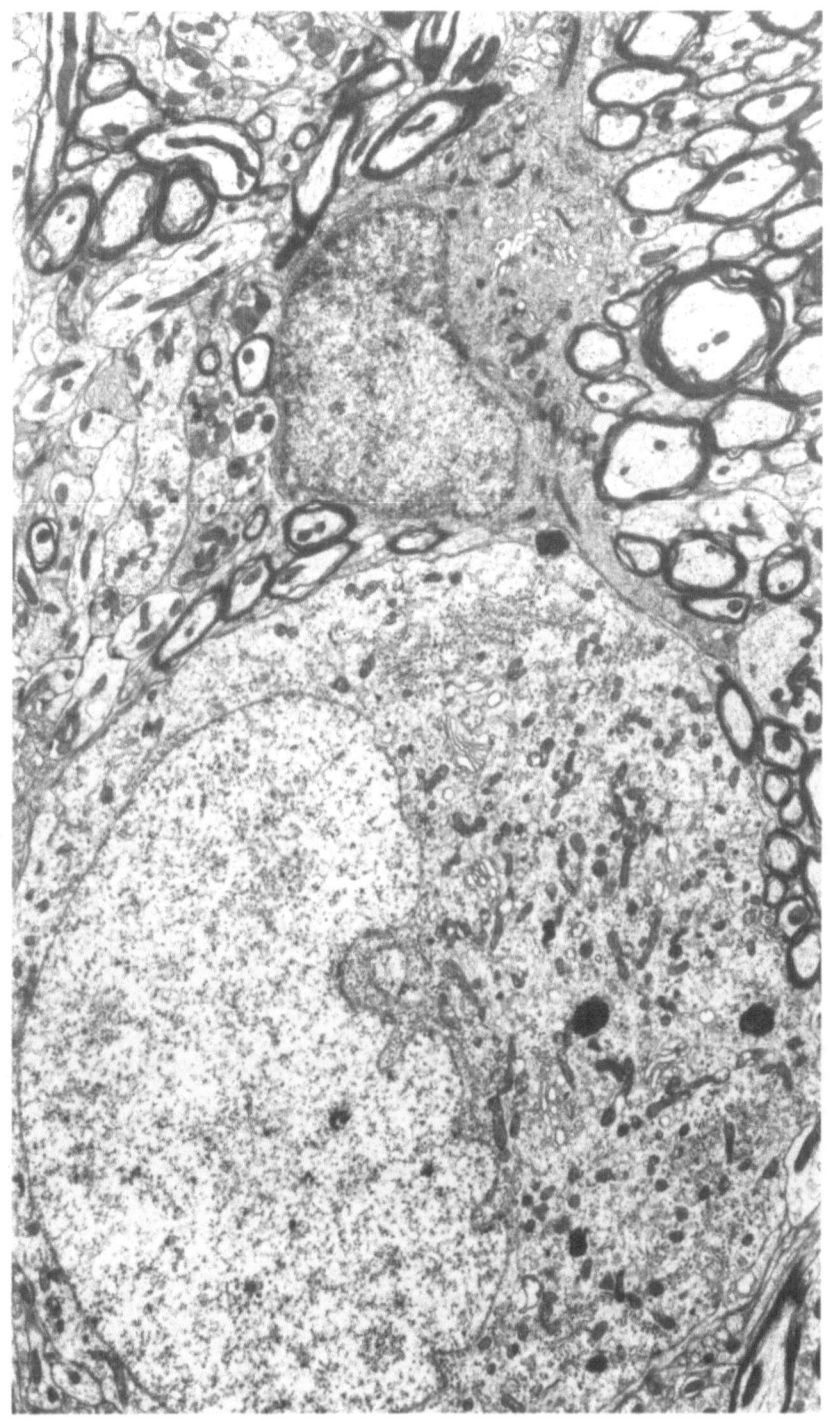

Fig. 31. Oligodendrocyte and its process in a perineuronal-perifascicular position (*arrow*). Ground squirrel. ×7,130

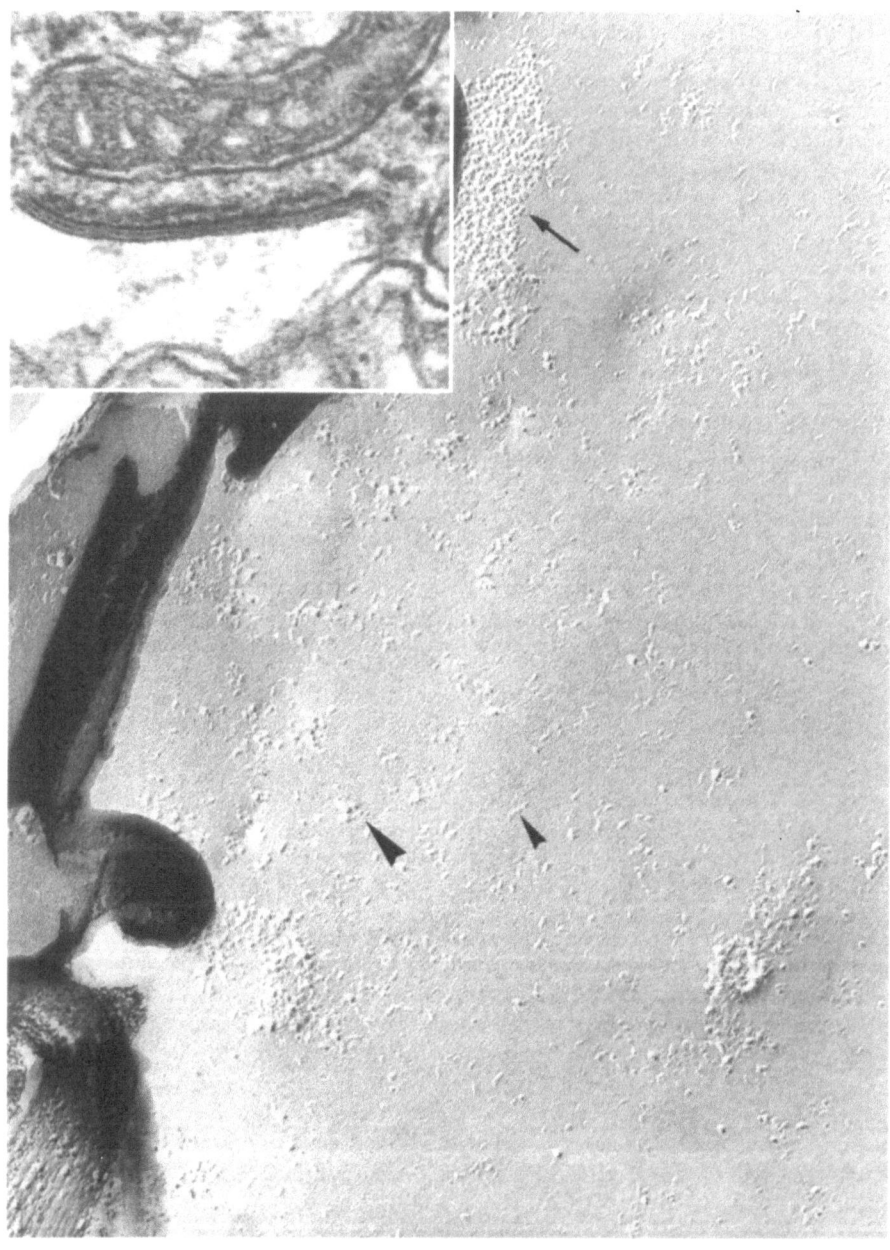

Fig. 32. Oligodendrocytic membrane with a heterogeneous population of globular (*large arrowhead*) and elongated (*small arrowhead*) intramembranous particles on the P-face. Cluster of gap junction particles (*arrow*). Cat. ×41.400. Inset: Astro-oligodendrocytic gap junction in a thin-sectioned preparation. Pigeon. ×34,000

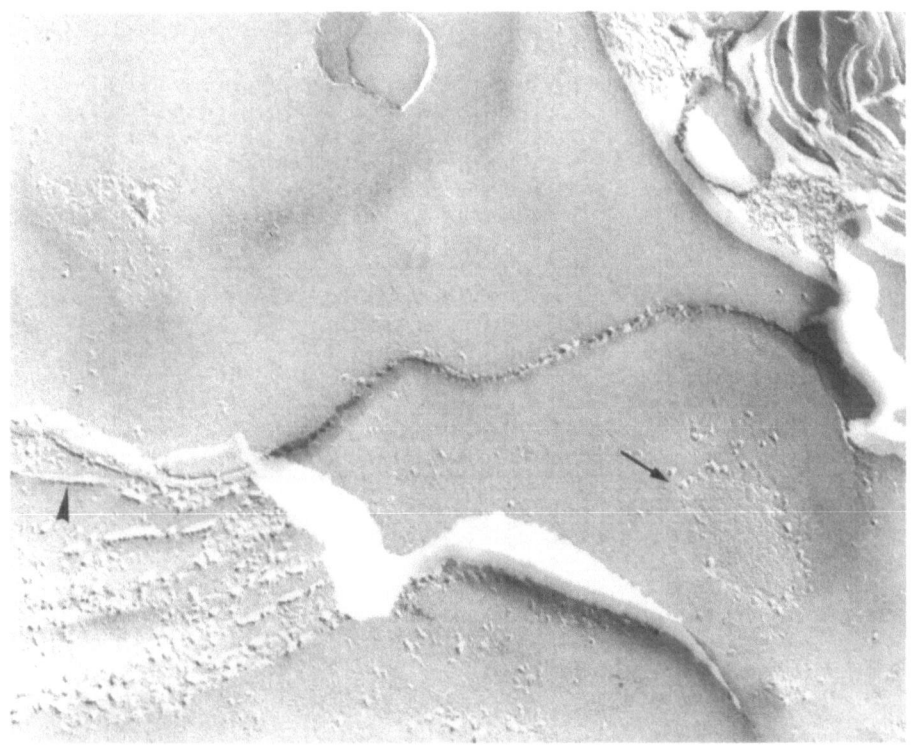

Fig. 33. Parallel ridges of intramembranous particles (*arrowhead*) on the oligodendrocytic P-face characterize the tight junction. In the vicinity of the tight junction, the gap junction E-face with arrays of pits (*arrow*) is exposed. Cat. ×41,400

3.2.2.6
Neuronal Plasticity and Spontaneous Degeneration in the Inferior Olivary Complex

Neurons, dendrites, and axons in the inferior olivary complex in every species, and all normal animals examined, contain the usual organelles, but examples of altered or degenerating profiles have been noticed. These unusual elements are dispersed among well-preserved neurons, dendrites, and axons.

The nucleus of some olivary neurons in the rat, ground squirrel, and cat include intranuclear filamentous inclusion, which can be extremely long in the cat and of a very special form in the ground squirrel. In the pigeon, only granulofilamentous inclusions are present. In hibernating ground squirrels, some neuronal nuclei contain filamentous inclusions in combination with membrane-bound inclusions (Fig. 34). The membrane-bound inclusions are composed of a granulofilamentous meshwork with various density aspects. These inclusions are also present in the dendritic profiles and axon terminals. Acidic phosphatase reaction product is located in these membrane-bound inclusions (Fig. 35). In the inferior olivary complex of adult rats, some large axonal profiles are alternated. They contain clear synaptic vesicles and membrane-bound granulofilamentous inclusions. Degenerating changes are also present.

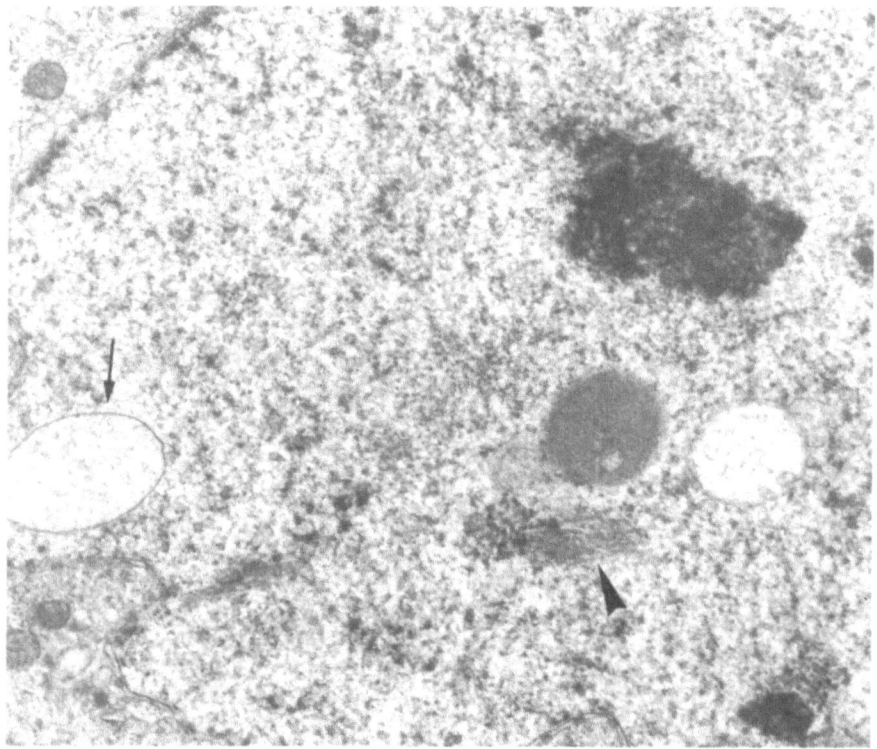

Fig. 34. Filamentous inclusions (*arrowhead*) in combination with membrane-limited inclusions (*arrow*). Ground squirrel. ×16,560

The perikarya of some olivary neurons in adult rats contain membrane-bound inclusions filled with filaments and nucleolus-like inclusions or nematosomes. In the rat olivary neurons, nematosomes are compact, small, and round in shape. Nematosomes are also present in the inferior olivary complex of ground squirrels, especially in females during the spring. These cytoplasmic inclusions exhibit a wide range of variations in size and shape. Nemsatosomes in the ground squirrel olivary neurons are composed of electron-dense strands 40–60 nm in diameter and surrounded by filaments (Fig. 36). These cytoplasmic inclusions are not marked with reaction products for acidic phosphatase (Fig. 36, inset). Nematosomes have not be observed in the cat and pigeon inferior olivary complex.

Cytoplasmic inclusions containing cisternal and tubular elements in combination with electron-dense material are found in olivary neurons of non-hibernating ground squirrels. These complex convolutions are in close association with cisterns of the granular endoplasmic reticulum, and ribosomes are associated with the external cisterns of these inclusions. Acidic phosphatase reaction product is detected on the surface of these cisternal or tubular profiles and is diffusely distributed on dense material between them. Such complex convolutions are also present in hypertrophied dendritic profiles of the rat inferior olivary complex. Some normal olivary neurons and dendrites in the cat exhibit submembranous dense bodies in apposition to the dendrites, while no specializations are involved.

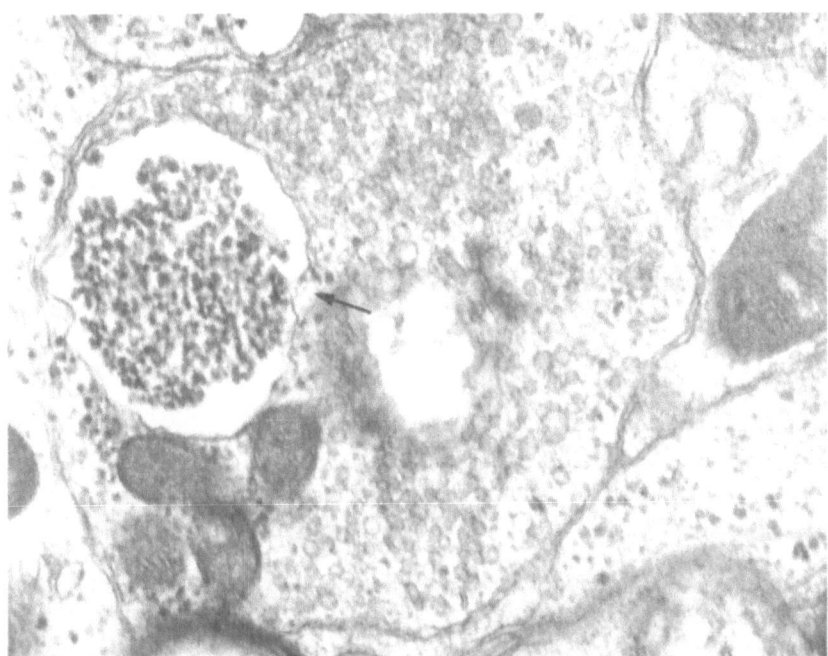

Fig. 35. Acidic phosphatase-reactive membrane-limited inclusion (*arrow*) in an altered axon terminal. Ground squirrel. ×56,700

In the normal inferior olivary complex of all animals examined, spheroids and different types of axonal alternations are present, especially in the ground squirrel. Acidic phosphatase reaction product is present in these spheroids. Electron-opaque degenerating axon terminals are also observed. Altered or degenerate axon terminals, degenerate myelinated axons, and other neuronal processes are encapsulated within a concentric lamellar sheath, which is composed of astrocytic lamellae. These multilamellar astrocytic capsules are most common in the inferior olivary complex of the ground squirrel. The lamellae composing multilamellar capsules vary in number and are occasionally found in continuity with astrocytes (Fig. 37). In some cases, the reactive microglial cells swallow up these multilamellar capsules surrounding masses of degenerate debris (Fig. 38). In freeze-etch replicas, reactive astrocytic membranes contain a large number of assemblies on the P-face and complementary pits on the E-face.

In the neuropil of the inferior olivary complex, especially of the ground squirrel, free postsynaptic densities without presynaptic specializations are presented. They are in apposition with neuronal somata, astrocytic lamellae, oligodendrocytes, or myelinated sheaths. Some of the free postsynaptic densities are part of the crest synaptic specialization in apposition with the myelin sheath (Fig. 39).

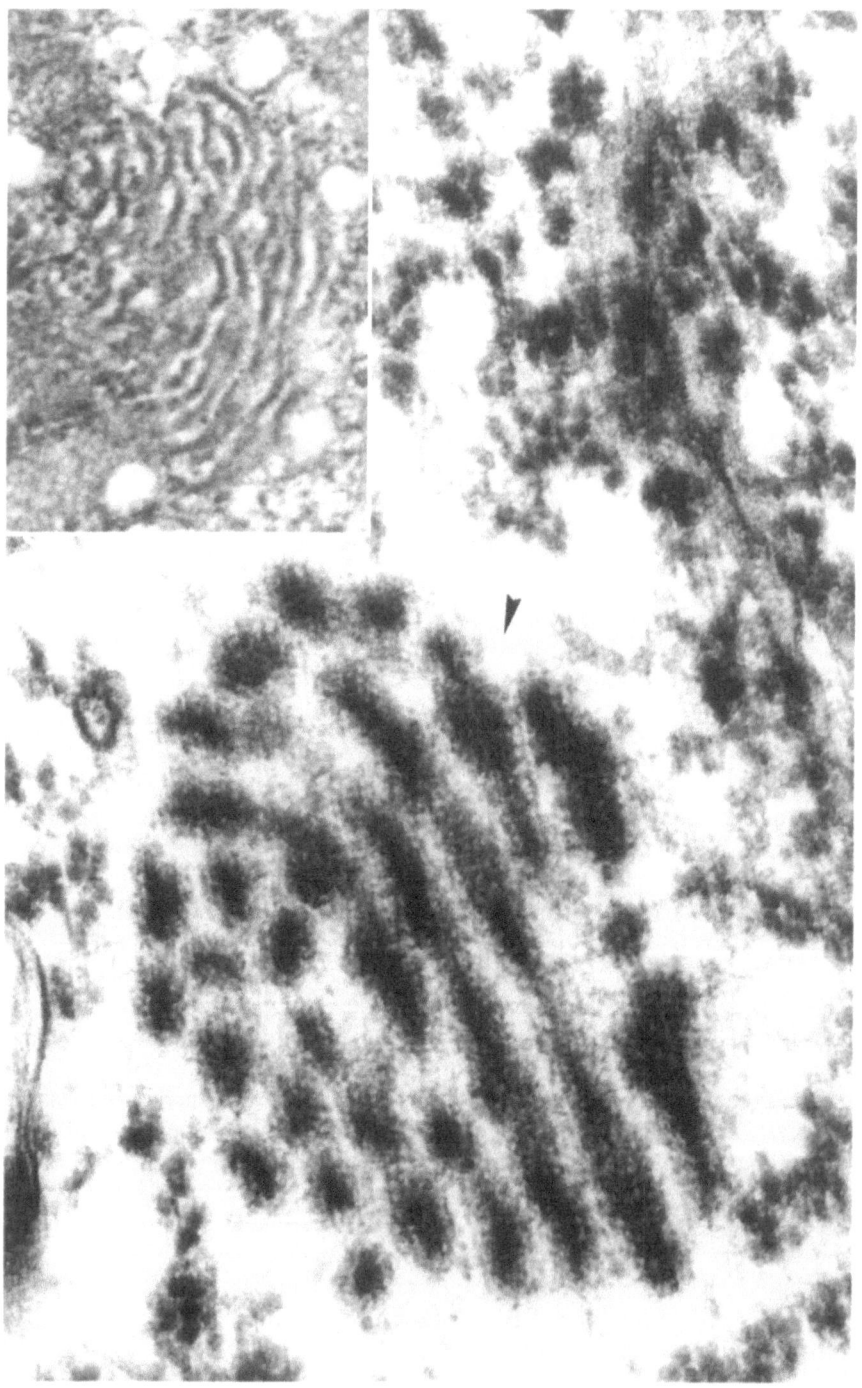

Fig. 36. Nematosome (*arrowhead*) in the perikarya of the female ground squirrel. ×90,000. Inset: Acidic phosphatase negative nematosome. ×14,300

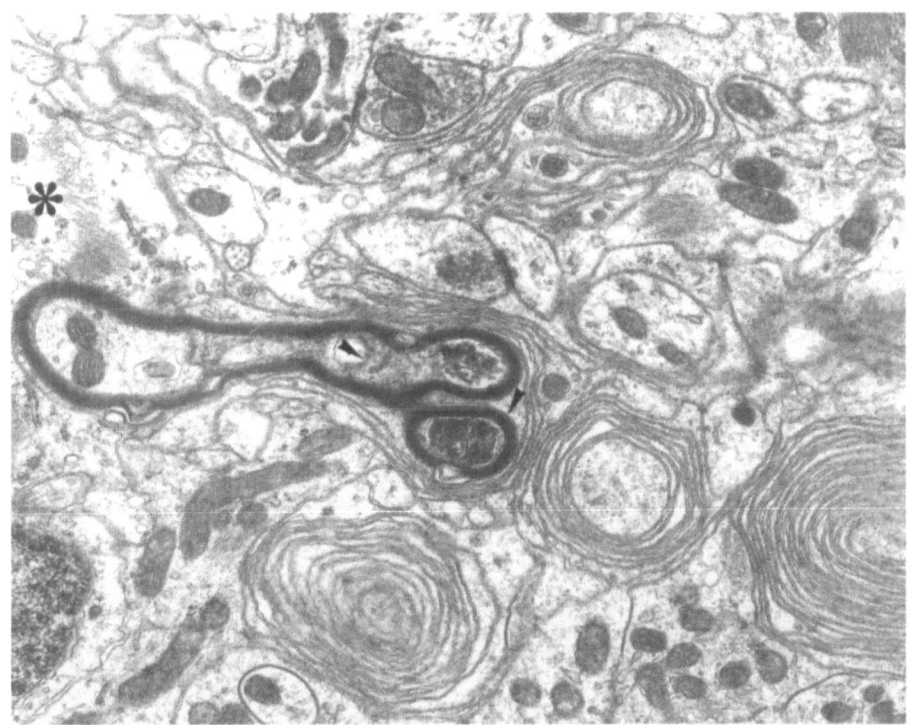

Fig. 37. The degenerate myelinated axons (*arrowheads*) are surrounded by astrocytic multilamellar capsule in succession with a reactive astrocyte (*asterisk*). Ground squirrel. ×13,800

4
Discussion

The inferior olivary nucleus develops from the same rostro-dorsal part of the rhombencephalic alar plate as the cerebellum, but migrates to the ventral parts of the rhombencephalon (reviewed in Nieuwenhuys et al. 1998). The olivary neurons are generated dorsally in the neuroepithelium of the rhombic lip of the fourth ventricle and migrate ventrally following submarginal and marginal streams (His 1890). The rat inferior olivary neurons are generated on embryonic day 12 (E12) in the medial accessory olive, on E13 in the dorsal parts of the dorsal accessory olive and ventral parts of the principal olive, and form a comma-shaped mass of cells until E18 (Bourrat and Sotelo 1988; 1990a,b). The parcellation of the inferior olivary complex is in close relation with the parcellation of the cerebellum during the development and maturation of the olivocerebellar connections (Robertson and Stotler 1974; Sotelo et al. 1984; Bourrat and Sotelo 1991; Landis et al. 1989; Heckroth et al. 1990; Sotelo and Wassef 1991; Armengol and Lopes-Raman 1996; Sotelo and Chadotal 1997.

The present correlated light- and electron-microscopic study represents an attempt to analyze morphological characteristics of neurons and glial cells in the inferior olivary complex of different representatives of submammalian and mammalian vertebrates, including humans.

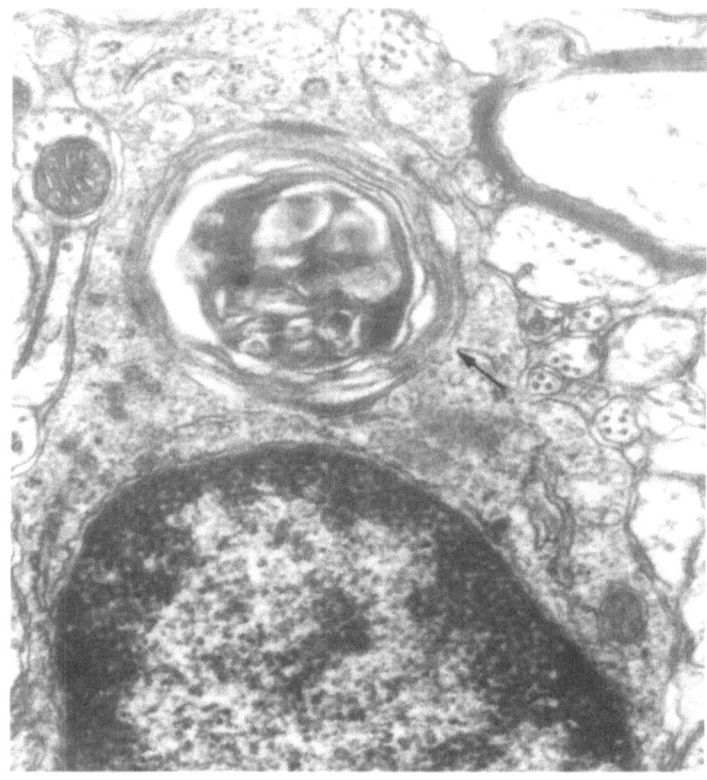

Fig. 38. The multilamellar astrocytic capsule surrounding degenerated debris (*arrow*) is embedded in the reactive microglial cell. Ground squirrel. ×35,100

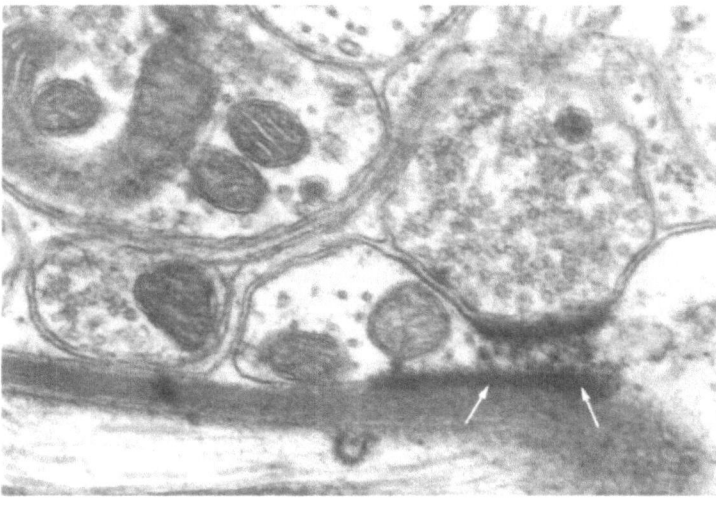

Fig. 39. The free postsynaptic density (*white arrows*) of the crest synaptic specialization in apposition with the myelinated axon. Ground squirrel. ×34,500

4.1
Light Microscopy

4.1.1
Topography and Cell Types of the Inferior Olivary Complex

Homologues of the inferior olivary complex in mammals, including humans, are present in submammalian vertebrates (Kooy 1917; Papez 1929; Ariens Kappers et al.1936; Whitworth and Haines 1986a). The topography of the carp inferior olivary complex, as described in the present Nissl-stained material, confirms previous data on the fish inferior olivary complex. Kooy (1917) made wax reconstructions of the inferior olivary complex in various fishes (Elasmobranchii, Teleostei, and Dipnoi). Because of a similar topography, this nuclear mass is regarded as homologous to the mammalian medial accessory olive (Kooy 1917). In the caudal rhombencephalon in the reedfish (*Erpetoichthys calaboricus*), an elongated group of small (11 μm) spherical neurons was described (Nieuwenhuys and Oey 1983). A group of small ovoid olivary cells was labeled in the teleost catfish (*Ictalurus mebulosis, Ictalurus natalis, Ictalurus punctatus*) after injections of horseradish peroxidase in the molecular layer of the contralateral cerebellar hemisphere (Finger 1978).

The analysis of our Nissl-stained material shows that morphological characteristics of the clusters of olivary neurons, which are situated in mediobasal parts of the caudal rhombencephalon in the frog (*Rana temporaria),* lizard (*Lacerta muralis),* and tortoise *(Tesdudo graeca),* are similar to those described by us in the fish inferior olivary complex. Homologue cell groups are defined as olivary neurons in amphibians and reptiles (see page 1, note 1: "The olive of Amphibia and Reptiles is too little circumscript to be drawn or reconstructed", Kooy 1917; Shanklin 1930; Bozhilova 1976). This cytoarchitectural group of olivary neurons is also identified on the base of physiological (Cochran and Hackett 1977) and anatomical tract-tracing studies (Künzle 1983; Grover and Grüsser-Cornehls 1984; Bangma and ten Donkelaar 1982; Künzle and Wiklund 1982; Wilczynski 1982; Künzle 1985; Van der Linde and ten Donkelaar 1987; Van der Linde et al. 1990).

In birds, the inferior olivary complex is composed of a large and more compact dorsal lamella and a smaller ventral lamella, which are rostrally connected at the medial angle (Williams 1909; Yoshimura 1909; Kooy 1917; Vogt-Nelsen 1954; Bozhilova-Pastirova et al. 1989a).

The inferior olivary complex in mammals, including humans, is composed of three topographically distinct subdivisions: medial and dorsal accessory olives, and a principal olive between them. Kooy (1917) made wax reconstructions of nearly all orders of Mammalia: Monotremata, Marsupialia, Insectivora, Chiroptera, Xenarthra, Rodentia, Carnivora, Cetacea, Perissodactyla, Artiodactyla, Proboscidea, Sirenia, Prosimi, Simia, and humans. Six principal subdivisions in the medial accessory olive, medial and lateral parts of the dorsal accessory olive, a dorsal lamella, a ventral lamella, and a lateral band between both principal olive lamellae have been described in the mammalian inferior olivary complex for a variety of mammalian forms (reviewed in Whitworth and Haines 1986a).

The topography of the inferior olivary complex in the rat, cat, and human in the present material confirms previous observations (Brodal 1940; Olszewski and Baxter 1954; Braak 1970; Gwyn et al. 1977). Moreover, this study represents the first attempt to analyze the topography of the inferior olivary complex of the ground squirrel (*Citellus citellus L.*). The caudal pole of this complex is formed by group b and in a rostral direction groups a, b, and c are present at the short distance, as is found in many mammalian species (reviewed in Whitworth and Haines 1986a). The dorsomedial part of the medial accessory olive is occupied by the dorsal cap, and the nucleus β appears most medially as was described in the cat (Brodal 1940; Walberg 1956) and the rat (Gwyn et al 1977). The ventrolateral outgrowth fuses with the principal olive and lies in close apposition to the dorsal accessory olive (in the cat) (Walberg 1956; Taber 1961). The dorsomedial cell column appears as a part of the rostral medial accessory olive, as in the cat (Walberg 1956; Taber 1961), and does not fuse with the principal olive in the manner similar to that found in the rat (Gwyn et al. 1977). The dorsal accessory olive in the ground squirrel has caudal-to-rostral extensions as in the cat (Walberg 1956), rhesus monkey (Bowman and Sladek 1973), and squirrel monkey (Rutherford and Gwyn 1980; Whitworth and Haines 1986b). It fuses with the ventral lamella of the principal olive as in the cat and rhesus monkey (Walberg 1956; Taber 1961; Bowman and Sladek 1973). The so-called tail portion of the rat and guinea pig dorsal accessory olive (Breazile 1967; Gwyn et al. 1977) is absent in the inferior olivary complex of the ground squirrel. The infoldings of the ventral and dorsal lamellae of the principal olive, like the primate olive (Bowman and Sladek 1973; Rutherford and Gwyn 1980; Whitworth and Haines 1986b) and human olive (Kooy 1917; Olszewski and Baxter 1954; Braak 1970), are not formed. The dorsal accessory olive and the principal olive fuse at the rostral pole as found in the opossum (Martin et al. 1975), cat (Brodal 1940; Walberg, 1956), and rhesus monkey (Bowman and Sladek 1973).

The cytological characteristics of the olivary neurons are studied in different submammalian vertebrates. Historically, Kooy (1917) identified the bird olivary neurons as pear- or spindle-shaped and rather small, though larger than in fish. Reticular neurons in the raphe of the pigeon (*Columba livia*) are described as large in size with a sharp, polygonal, long-stretched form. In thionin-stained sections, Vogt-Nilson (1954) also recognized polygonal, rounded, elongated, and spindle-shaped olivary neurons, and concluded that the largest avian olives could often contain the largest cells. The diameter of the olivary neurons in the mature chicken (P35) ranges between 16 and 21 μm (Furber 1984) as this is also found in the pigeon (Bozhilova-Pastirova et al. 1989a).

In this study we present a description of the olivary neurons in the pigeon (*Columba livia*), which are essentially similar everywhere in the inferior olivary complex. In addition, large reticular-like neurons are observed in periolivary and intraolivary positions.

Most authors have identified the neurons of the mammalian inferior complex in different species, including humans, as rounded to oval in form and medium in size (Scheibel and Scheibel 1955; Bowman and King 1973; Sotelo at al. 1974; Gwyn et al. 1977; Foster and Peterson 1986; Bozhilova-Pastirova 1990c)

The previous quantitative studies have described the shape, size, and packing densities of the olivary neurons in the bat, rat, guinea pig, cat, and human (Moatamed 1966; Escobar et al. 1968; Schild 1970; Mlonyeni 1973; Foster and Peterson 1986). Scheibel and Scheibel (1955) provide an exception examining the average number of cells per unit volume (40 μm thick sections/200 μm² area) in the inferior olivary complex of the cat, monkey, and human.

Our report concerns results obtained from Nissl-stained preparations about measurements of the mean area with the mean maximal and minimal diameter and mean neuronal density in the inferior olivary complex of the carp, pigeon, ground squirrel, cat, and human. These results are displayed in Figs. 2–4. The quantitative comparison with these parameters in the inferior olivary complex indicates a decrease in neuronal density from carp to human (Fig. 5), which seems to be concomitant of an increase in mean neuronal cell body area towards the human. In the mammalian inferior olivary complex, ratios between small and large neurons are different from those in the submammalian inferior olivary complex (Figs. 2–4). The shape of the neuronal perikarya is represented by the elongation index, which is the ratio of the maximal diameter to the minimal diameter. Most of the olivary neurons in submammalian species have slightly asymmetrical cell bodies with a ratio greater than 1:1.2. Almost rounded cell bodies are present in olivary neurons of the ground squirrel, cat, and human with an elongation index of 1:1.2.

The population of reticular neurons is located in a periolivary position (Scheibel and Scheibel 1955; Bowman and King 1973; Sotelo et al. 1974; Gwyn et al. 1977; Rutherford and Gwyn 1980; Foster and Peterson 1986; Bozhilova-Pastirova 1990c). This population is generally magnocellular. In humans, some neurons in the caudal parts of the medial accessory olive indeed have the same morphological characteristics of the reticular neurons described by Scheibel and Scheibel (1955).

4.1.2
Dendritic Morphology of Inferior Olivary Neurons

Ramón y Cajal (1909) was the first to obtain complete impregnation of the dendritic tree of the olivary neurons in human material. He described only one type of olivary neuron of which the dendrites divide and redivide and after folding into themselves have the appearance of "a ball of wool". The population analysis in the inferior olivary complex is based on recognizing criteria, which describe the arrangement of the dendritic tree and other dendritic features. The cell types described by these criteria are referred to as two types. The first type, with unramified dendrites which run away from the cell body, is located predominantly in caudal portions of both accessory olives in the rat, guinea pig, ground squirrel, cat, and monkey (Scheibel and Scheibel, 1955; Scheibel et al. 1956; Gwyn et al. 1977; Bozhilova and Ovtscharoff 1979; Rhuterford and Gwyn 1980; Foster and Peterson 1986; Bozhilova-Pastirova 1990c; Ruigrok et al. 1990a; Ruigrok and De Zeeuw 1993; Bozhilova-Pastirova and Ovtscharoff 1995a). The second type has extensively ramified dendrites, which turn back towards the cell body giving the appearance of a coiled ball (Ramón y Cajal 1909; Scheibel and Scheibel 1955; Sotelo et al. 1974; Gwyn et al. 1977; Bozhilova and Ovtscharoff 1979; Rhuterford and Gwyn 1980; Foster and Peterson 1986; Bozhilova-Pastirova 1990c; Ruigrok et al. 1990a, 1993; Bozhilova-Pastirova and Ovtscharoff 1995a). Foster and Peterson (1986) claimed that the second type can be subdivided in type II-a (a transitional form between the relatively unbranched type I) and type II-b (the tightly wound type II-a). According to us, when the shape of the dendritic field is taken into account, a third type of mammalian olivary neuron might be accepted – marginally located neurons with dendrites that project into the nucleus. These neurons commented on by Ramón y Cajal (1909), but not subdivided, are present in mammalian species (Scheibel and Scheibel 1955; Bowman and King 1973; Sotelo et al. 1974) including humans (Fig. 7d). The reticular-like type of neuron with a peripheral position, or placed among the typical olivary neurons, might be accepted as an additional cell type.

The olivary neurons in the mammalian inferior olivary complex can be classified as a moderate spiny type possessing so-called spine crowned appendages (Bowman and King 1973; Sotelo et al. 1974; Gwyn et al. 1977; Bozhilova and Ovtscharoff 1979; Rhuterford and Gwyn 1980; Bozhilova-Pastirova 1990c).

In the submammalian inferior olivary complex, the first type of neuron with dendrites that radiate away from the cell body is identified in the chicken (Furber 1984) and the pigeon (Bozhilova-Pastirova et al. 1989a; Bozhilova-Pastirova and Ovtscharoff 1995a). In the intermediate part of the dorsal lamella in the pigeon we found neurons with a dendritic pattern close to the second type of mammalian inferior olivary complex (Bozhilova-Pastirova et al. 1989a; Bozhilova-Pastirova and Ovtscharoff 1995a). These rostral parts of the dorsal lamella of the avian inferior olivary complex are accepted as homologues of the principal olive (Furber 1983).

The present study adds a quantification of the dendritic field area of all types of olivary neurons in submammalian and mammalian species examined, including humans. The dendritic parameters are usually examined by different Golgi impregnation methods, but in the guinea pig (Foster and Peterson 1986) and the cat (Ruigrok et al. 1990a) olivary neurons are visualized with horseradish peroxidase (HRP) intracellular injection. Thus, in the phylogeny, the inferior olivary neurons in submammalian species are densely packed and dendritic fields are more overlapping. A comparison can be made between quantification of the dendritic field in the chicken (Furber 1984) and our measurements in the pigeon inferior olivary complex, which are in a good agreement. In the carp inferior olivary complex the dendritic field is relatively large in comparison with the greatest neuronal density. This study expands previous observations (Scheibel and Scheibel 1955) on the dendritic morphology in different mammalian species including humans. It indicates a trend for humans to have a smaller neuronal density and smaller dendritic field area, which seems to be concomitant with the increase in neuronal area. This reflects an increase in the dendritic field restriction as a base for the divergence of multiple synaptic input and for synchronous firing in the mammalian inferior olivary complex (Scheibel and Scheibel 1955; Sotelo et al. 1974; Llinás et al 1974). In the local circuitry of the inferior olivary complex, dendrites of the periolivary (Scheibel and Scheibel 1955; Bishop and King 1986; Bozhilova-Pastirova and Ovtscharoff 1995a) or intraolivary-located reticular neurons also participate. The dendritic field area was measured in the pigeon, ground squirrel and cat, and was greater in comparison with the typical olivary neurons.

Dendrites of mammalian olivary neurons usually remain within the confines of the subnuclei (Scheibel and Scheibel 1955; Scheibel et al. 1956), but in some cases Golgi-impregnated dendrites of olivary neurons in the ground squirrel principal olive and dorsal accessory olive overlap. However, there are certain examples that the dendrites of some neurons in the ground squirrel dorsal accessory olive tend to extend across the middle into the contralateral olive as it is observed in the rat (De Zeeuw et al 1996). It was accepted that the olivary compartments should be considered as dynamic states and not as an anatomical entity (Llinás and Sasaki 1989).

The axons extend from either the cell body of the first type of olivary neuron or from proximal dendrites of the second and third types of olivary neurons without any differences in morphological characteristics. These findings are in agreement with the observations on the intracellular HRP-stained neurons (De Zeeuw et al. 1990c; Ruigrok and De Zeeuw 1993).

4.1.3
GABA and Parvalbumin Immunoreactivity
in the Inferior Olivary Complex

Aspartate and glutamate are thought to be the main transmitters of the olivocerebellar projection neurons (Wiklund et al. 1982; Aoki et al. 1987). Recent immunocytochemical studies have revealed that certain subpopulations of olivocerebellar neurons express a variety of peptides and messenger RNA (for references, see Ueyama et al. 1994; Marcos et al. 1994). Peptides are also visualized in axon terminals and varicosities of the inferior olivary neuropil that probably could be colocalized with neurotransmitters such as acetylcholine, catecholamines, or GABA (Gregg and Bishop 1997). Calbidin-immunoreactive olivary neurons were observed at E16 in adult rats, while olivary neurons immunoreactive for parvalbumin and calcitonin gene-related peptide were transiently observed during development (Wassef et al. 1992a,b). It was interesting to understand whether parvalbumin immunoreactivity could be detected after the so-called creeper stage in cerebellar climbing fiber synaptogenesis (Chedotal and Sotelo 1993). Our results indicate that GABA- and parvalbumin-immmunoreactive neurons are present in the dorsal cap and small parts of nucleus β of the rat medial accessory olive at P20. We have also determined the ratio of parvalbumin-immunoreactive to GABA-immunoreactive neurons. Our data concerning neuronal counts in male and female rats suggest significant sex differences in the inferior olivary complex. Females have a greater number of GABA- and parvalbumin-immunoreactive neurons than males (Table 4), as it has been found in different regions of the rat brain (Ovtscharoff et al. 1992; Stefanova et al. 1997a,b). The ratio of GABA-immunoreactive:parvalbumin-immunoreactive olivary neurons in females (2.0066:1) is similar to that in males (2.66:1).

In the inferior olivary complex, a subpopulation of GABA-immunoreactive neurons is also present (Nelson and Mugnaini 1988; Walberg and Ottersen 1989; Fredette et al. 1992). According to Fredette et al. (1992), in the rat, rabbit, cat, rhesus monkey, and human, GABAergic neurons are subdivided into three categories: periolivary neurons in the gray matter and the white matter around the nuclear complex, interolivary neurons between the subnuclei of the inferior olivary complex, and intraolivary neurons. Our results are in a good agreement with these data. GABA-immunoreactive terminals densely supply the mammalian inferior olivary complex (Sotelo et al. 1986; Gotow and Sotelo 1987; Nelson et al. 1989). These terminals differ in size and staining intensities and our results about distribution patterns and density of the GABA-immunoreactive terminals are in agreement with previous data (Nelson et al. 1989). GABAergic projections from the cerebellar, vestibular, reticular, pretectal, and prepositus hypoglossi nuclei to the inferior olivary complex (Bishop 1984; Nelson et al. 1984; 1986; Angaut and Sotelo 1987; Nelson and Mugnaini 1989; Fredette and Mugnaini 1991; De Zeeuw et al. 1993; 1994; Barmack et al. 1998) are topographically organized. These projections modulate the pattern generation properties of neuronal assemblies in the olivocerebellar system (Lang et al. 1996). According to Sotelo et al. (1986), the interpretation of Perez de la Mora et al. (1981) that larger and strongly immunoreactive terminals belong to the projection neurons, while the punctate and weakly to moderately immunoreactive terminals are characteristic of GABA innervation by interneurons, cannot be accepted as criteria for identification. Anatomical (Sotelo and Arsenio-Nunes 1976; Angaut et al. 1985; Sotelo et al. 1986) and physiological studies (Grill 1970; Llinás et al. 1974; Llinás and Yarom 1981a,b) indicate that most, if not all, olivary neurons are projecting neurons.

Regarding the group of GABA-immunoreactive olivary neurons, their projections and physiological implications are not well known. On the other hand, it is known that GABAergic neurons can be either local circuitry interneurons or projection neurons in various brain regions (for references, see Mugnaini and Oertel 1985). These data allow us only to speculate that GABA-immunoreactive olivary neurons are projecting neurons. In addition, parvalbumin-immunoreactive olivary neurons could be accepted as a subpopulation of projection GABAergic neurons as it is demonstrated in the septo-hippocampal area (Freund 1989). Furthermore, the number of GABA-immunoreactive neurons is higher than that of parvalbumin-immunoreactive neurons in the dorsal cap and nucleus β of the medial accessory olive of male and female rats. These sex differences are significant during postnatal development.

4.2
Electron Microscopy

Walberg (1963, 1964) carried out the first investigation on the ultrastructure of the neurons and glial cells in the cat inferior olivary complex. The present study provides a description of the fine structure of the perikarya and related neuropil in the pigeon, rat, ground squirrel, and cat.

In general, the somatic morphology of the ground squirrel resembles those reported in the opossum, rat, cat, and monkey (Bowman and King 1973; Sotelo et al. 1974; Gwyn et al. 1977; King 1980; Rutherford and Gwyn 1980; Bozhilova-Pastirova 1990b; Ruigrok and De Zeeuw 1993). Since virtually all olivary neurons examined in this study are medium-sized and have similar somatic ultrastructure, it is plausible to suggest that these neurons are projection neurons. This assumption is consistent with the results after axotomy of olivary neurons in the inferior cerebellar peduncle of immature rats (Sotelo and Aresenio-Nunes 1976; Angaut et al. 1985), following intracellular recording in the inferior olivary complex (Grill 1970; Llinás et al. 1974) and in vitro slices (Llinás and Yerom 1981a,b), which indicated that most, if not all, olivary neurons are projecting neurons. According to Rutherford and Gwyn (1980), two types of olivary neurons are present in the dorsal accessory olive of the squirrel monkey, but the second type, made up of small (13 µm in diameter) neurons with a large deeply indented nucleus and thin rim of cytoplasm, is very uncommon. Classical observations (Vincenzi 1886–1887; Ramón y Cajal 1911) have illustrated neurons with recurrent collaterals in Golgi-impregnated human inferior olivary complex, in newborn infants, and 4-day-old kittens. This could be accepted as a base that the same projecting neurons also serve as interneurons via their collaterals. However, this was not confirmed, as revealed recently in intracellularly HRP-labeled olivary neurons (King 1980; Foster and Peterson 1986; Ruigrok and De Zeeuw 1993). Reticular-like neurons are found in the intraolivary, interolivary, and periolivary position in the ground squirrel inferior olivary complex, and their dendrites and axon terminals probably participate in the local circuit as it has been demonstrated in the rat inferior olivary complex (Bishop 1984; Bishop and King 1986). Moreover, larger neurons in periolivary position are found to be GABAergic (Fredette et al. 1992; A. Bozhilova-Pastirova and W. Ovtscharoff, personal observations), and might be a source of GABA-immunoreactive axon terminals to the inferior olivary complex.

The pigeon olivary neurons are smaller in size but possess typical ultrastructural characteristics as those in the mammalian olivary complex (Bozhilova-Pastirova et al. 1989a). In general, the fine structure of the pigeon inferior olivary nucleus resembles that reported in different mammalian species (Bowman and King 1973; Sotelo et al. 1974; Gwyn et al. 1977; Rutherford and Gwyn 1980; Bozhilova-Pastirova 1990b). Ultrastructural characteristics of the reticular-like neurons between typical olivary neurons and around the lamellae in the pigeon inferior olivary complex are reported in the present study. These periolivary reticular-like neurons send their dendrites into the pigeon inferior olivary complex (Bozhilova-Pastirova and Ovtscharoff 1995a). At present there are no data indicating that these neurons are involved in the olivary circuit as it has been seen in the rat inferior olivary complex (Bishop and King 1986). Moreover, whether the pigeon's interolivary and periolivary reticular-like neurons are GABAergic or not remains to be determined.

The plasma membrane of neurons and proximal dendrites in the inferior olivary complex are covered with glial cells and their processes (Walberg 1963, 1964; Sotelo et al. 1974). However, in many cases extensive membrane appositions are present, which occur between somata of the olivary neurons in the ground squirrel, soma and dendrites, as it has been noticed in the cat (Sotelo et al. 1974). A few puncta adhaerentia, but no gap junctions, have been observed at these sites in our thin-section preparations. Some olivary neurons in caudal portions of the medial accessory olive of the ground squirrel possess cilia with a modified pattern. It may be a vestigial structure as in other parts of the nervous system (Dahl 1963; Del Cerro and Snider 1967; Peters et al. 1976; Lafarga et al. 1980).

4.2.1
Synaptic Relationships

Analysis of the ultrathin sections in the pigeon, ground squirrel, rat, and cat inferior olivary complex revealed that one type of axon terminal contains predominantly rounded vesicles while the second type contains pleomorphic vesicles. In addition, both types contain dense-core vesicles. Axon terminals often form more than one synapse of intermediate, asymmetric, or symmetric type. The majority of synapses formed by the second type of axon terminal were symmetrical and their source is very likely to be GABAergic cerebello-olivary projections (for references, see Ruigrok and De Zeeuw 1993). The sites of termination shown by the first type of axon terminal forming asymmetrical synapses were similar to those of non-GABAergic mesodiencephalo-olivary fibers (De Zeeuw et al. 1988; 1989a,b). Sometimes asymmetric synaptic contact zones with subsynaptic dense bodies and crest synapses with somatic and dendritic spines in extra- or glomerular position are found. The crest synapses have been described in the rabbit (Mizuno et al. 1974), monkey (Rutherford and Gwyn 1980), and rat (Sotelo et al. 1986; De Zeeuw et al. 1989a; 1993). One or both types of axon terminals are GABAergic (De Zeeuw et al. 1989b) and may be of different origin (De Zeeuw et al. 1993). The axon terminals and boutons en passant contain a large number of dense-core vesicles. They do form not any specializations with other elements in their surrounding. These axon terminals and varicosities may represent serotoninergic boutons (Wiklund et al. 1981a,b; King et al. 1984). It is also possible that serotonin may be colocalized with GABA (De Zeeuw et al. 1989b) in the same terminal. Origin and distribution of serotonin-immunoreactive fibers and varicosities located in distinct olivary subnuclei were studied in different mammals (for references, see Bishop and Ho 1984, 1986). It has been indicated that these terminals are involved in the induction of the harmaline-generated tremor by a tonic inhibitory effect on the olivary neurons preventing harmaline (Sjölund et al. 1977). There is physiological evidence that serotonin increase the average firing rate of the olivary neurons, slows their oscillation frequency, and increases the coherence of their oscillators (Sugihara et al. 1995). In the neuropil of the ground squirrel inferior olivary complex we found dendritic profiles containing a mixture of clear rounded vesicles and dense-core vesicles close to the synaptic contact zones with two axon terminals (Fig. 12). Such types of dendritic profiles are described in the olivary neuropil of the rat (Gwyn et al. 1977) and opossum (King 1980), and in the dendrites of the marginal reticular neurons of the rat (Bishop and King 1986).

Complex synaptic junctions or glomeruli are present in the neuropil in all subnuclei of the inferior olivary complex in the ground squirrel as found in different mammalian species (Nemecek and Wolff 1969; Bowman and King 1973; Sotelo et al. 1974; King 1976; Gwyn et al. 1977; Bozhilova and Ovtscharoff 1979; Rutherford and Gwyn 1980; King 1980; De Zeeuw et al. 1989a, 1990a; 1996; Ruigrok and De Zeeuw 1993). In the course of our studies on the pigeon inferior olivary complex, we have observed synaptic arrangements with puncta adhaerentia between dendrite profiles in the central core, but their nature is not yet clear. Whether these contacts are adhesive or have a functional significance remains to be elucidated. So far we have not observed gap junctions between them. On the basis of these data, and the fact that the development of the dendro-dendritic gap junctions is a rather late process (between P10 and P15) in the rat inferior olivary complex (Bourrat and Sotelo 1983; Gotow and Sotelo 1987), it might be hypothesized that electrotonic transmission occurs later in phylogeny. An association of the gap junction and puncta adhaerentia was observed as it has been reported in the vertebrate nervous system (Sotelo and Llinás 1972; Bozhilova and Ovtscharoff 1979).

In the ground squirrel inferior olivary complex, the central core of the glomerulus is composed of up to 12 spine appendages. These profiles extend from dendrites in the extraglomerular position and gap junctions are formed between appendages of different dendrites. The spine appendages of Golgi-impregnated dendrites are in apposition with non-impregnated profiles in different complex synaptic domains. The results of our analysis correspond to those obtained from HRP intracellularly labeled olivary neurons (De Zeeuw et al 1990a; Ruigrok and De Zeeuw 1993).

The present study extends the observations on the synaptic organization in the mammalian and submammalian inferior olivary complex by analysis of the intramembranous structure of the postsynaptic membrane, presynaptic membrane, gap junctions, and puncta adhaerentia.

Data from freeze-etching replicas indicated that the E-face aggregates of intramembranous particles are freeze-etching equivalents of the asymmetrical synapses, which are usually excitatory (Landis et al. 1974; Landis and Reese 1974b). On the basis of physiological, morphological, and topographical findings, most non-GABAergic inputs to the rostral medial accessory olive deriving from the mesodiencephalic projections are probably excitatory (Bernardo and Foster 1986; Ruigrok and Voogd 1988; De Zeeuw et al. 1989b). The results for the particle packing density and size of the intramembranous particles within the postsynaptic E-face aggregates in the ground squirrel inferior olivary complex are comparable with the same results in the rat cerebellum and hippocampus (Harris and Landis 1986) or the rat sensorimotor cortex (Bozhilova-Pastirova and Ovtscharoff 1996a,b). We could not identify perforated aggregates at the postsynaptic E-face as seen in other regions of the central nervous system (Harris and Landis 1986; Surchev 1992; Bozhilova-Pastirova and Ovtscharoff 1996 a,b; 1999; Bozhilova-Pastirova 1998). Frequently, one axon terminal was found to make two or more synaptic contacts with one dendritic profile. In freeze-etching replicas, we found that the postsynaptic E-face aggregates are exposed near the presynaptic P-face modulations of the next synaptic contact zone (Fig. 20).

GABAergic terminals of cerebello- and vestibulo-olivary projections associated with gap junctions make symmetrical synapses, which are probably inhibitory (Nelson et al 1984; Angaut and Sotelo 1987, 1989; De Zeeuw and Koekkoek 1997) in all the subnuclei of the inferior olivary complex. In freeze-etching replicas the postsynaptic membrane of the inhibitory synapses is unspecialized (Landis and Reese 1974b; Landis et al. 1974). The synapses are unequivocally identified when the fracture process exposes simultaneously the presynaptic clusters of vesicles or part of the presynaptic P-face modulations and E-face of the unspecialized postsynaptic E-face. These synapses resemble the inhibitory synapses in the cerebellum (Landis and Reese 1974b; Harris and Landis 1986), the olfactory bulb (Landis et al. 1974) and the sensorimotor cortex (Bozhilova-Pastirova and Ovtscharoff 1996a,b; Bozhilova-Pastirova 1998) These membrane patterns may reflect differences in the proportions of the receptor subtype or postsynaptic proteins (Landis et al. 1987), which can be modified by activation of the Ca2+/calmodulin protein kinase II (Suzuki et al. 1992) and are visualized as intramembranous particles (Landis et al. 1987).

The presynaptic membrane of both excitatory and inhibitory synapses contains intramembranous particles preferentially associated with the P-face and a presynaptic modulation, i.e., dimples on the P-face and protuberances on the E-face. The mean size of the intramembranous particles at the synaptic contact zone on both faces of the presynaptic membrane is not significantly different, as found in the rat sensorimotor cortex (Bozhilova-Pastirova and Ovtscharoff 1996b; Bozhilova-Pastirova 1998). It has been suggested that the large intramembranous particles may represent ion channels (Pumplin et al. 1981) and presynaptic modulations are sites of synaptic vesicle exocytosis (Akert et al. 1972; Pfenninger et al. 1972; Miller and Heuser 1984).

The present study demonstrates the intramembranous characteristics of the dendro-dendritic puncta adhaerentia and gap junctions. Dendro-dendritic gap junctions are characterized by tightly packed homogeneous P-face particles in small aggregates, as found in the rat sensorimotor cortex (Bozhilova-Pastirova and Ovtscharoff 1995b). In some cases, gap junctions are large and intramembranous particle aggregates are loosely arranged with small islands of smooth membrane at the P-face. Complementary arrays of pits are present at the dendritic E-face across a narrowing of intercellular space and coated invaginations, as it is seen in thin sections.

The functional role of the neuronal gap junction is still a matter of debate. It has been reported that neurons may be synchronized by electrotonic coupling (Llinás et al. 1974). The dendritic lamellar bodies (De Zeeuw et al. 1995) are associated with dendro-dendritic gap junctions. These dendritic lamellar bodies are specifically labeled with antiserum (α12B/18) and start to occur postnatally between P10 and P15 (De Zeeuw et al. 1995; De Zeeuw and Koekkoek 1997). Their location is mediated by CLIP-115, a brain specific cytoplasmic linker protein (De Zeeuw et al. 1997a). It is known that gap junctions are capable of forming channels, which allow diffusion of cAMP, inositol triphosphate, and Ca+2. Neurotransmitters, voltage, pH, and Ca+2 can modulate the permeability (Saez et al. 1993).

Finally, the olivary glomerulus contains a gap junctional contact, which is characterized by aggregates of tightly packed intramembranous particles in a hexagonal pattern on the E-face of the dendritic appendage (Fig. 34). The fracture process exposes the unspecialized E-face of this dendritic appendage, the presynaptic P-face, and the organelles of the presynaptic cytoplasm with synaptic vesicles of the inhibitory synapse. The fracture plane intersected two parent dendritic profiles in connection with dendritic appendages and simultaneously exposed the unspecialized E-face of the adjacent postsynaptic dendrite in contact with the same presynaptic axon terminal. Synapses with an unspecialized postsynaptic membrane largely outnumber those with an E-face aggregate of intramembranous particles in the olivary glomeruli. Our results correspond to the distribution pattern of GABAergic and non-GABAergic terminals next to the dendrites coupled by gap junctions as defined by De Zeeuw et al. (1989a, 1990b).

In summary, our results of freeze-etching replicas demonstrate that in the olivary glomerulus, dendritic appendages connected by gap junctions make synapses. At the synaptic contact zone, the postsynaptic membrane has an intramembranous structure, either inhibitory or excitatory synapses, and is unequivocally identified in freeze-etching replicas. This is in keeping with light-, electron-microscopical, and electrophysiological and/or dye-coupling studies which demonstrated that electrotonically coupled olivary neurons (Llinás et al. 1974; Llinás and Yarom 1981a,b; 1986; Bernardo and Foster 1986; Llinás and Sasaki 1989; Sasaki et al. 1989) by dendro-dendritic gap junctions (Sotelo et al. 1974) are modulated by GABA (Sasaki and Llinás 1985; Lang et al. 1996) via presynaptic GABAergic axon terminals of cerebellar nuclei (Nelson et al. 1984; Sotelo et al. 1986; De Zeeuw et al. 1989a). It was proposed that the combined GABAergic cerebellar and excitatory non-GABAergic terminals regulate the electrotonic coupling between inferior olivary neurons and their firing frequency simultaneously (De Zeeuw et al. 1990b).

4.2.2
The Neuroglial Cells

Neuroglial cells in the mammalian inferior olivary complex were identified by Scheibel and Scheibel (1955) in Golgi-stained material as fibrous or protoplasmic astrocytes in the capsular region, along the fiber bundles in the gray matter and attached to the wall of the blood vesicles. They found orange-stained bodies representing the nuclei of oligodendrocytes. Walberg (1963, 1964) described ultrastructural details of the fibrous, protoplasmic astrocytes and oligodendrocytes. It was claimed that fibrous astrocytes might be the myelin-forming cells in the inferior olivary complex in the adult cat (Walberg 1964; Mugnaini and Walberg 1964). Astrocytes, their processes, and oligodendrocytes have a topographical organization as an immediate environment of the neurons and elements in the neuropil in the inferior olivary complex (Sotelo et al. 1974). During development, glial cells and immature neurons have similar characteristics, but can be identified by the presence of the glial cytoskeletal proteins S100 and GFAP (Cunningham et al. 1999).

In our Golgi-stained material we identify three varieties of astrocytes. The processes of the fibrous astrocytes are long, GFAP-immunoreactive, and located between bundles of myelinated fibers (Bozhilova-Pastirova et al. 1989d; Bozhilova-Pastirova et al. 1991b). The second type is made up of protoplasmic astrocytes with the variety of so-called velate astrocytes (Chan-Palay and Palay 1972; Palay and Chan-Palay 1974; Chan-Palay 1977; Bozhilova-Pastirova et al. 1991b). In thin sections, protoplasmic astrocytes have fewer intermediate filaments in the cytoplasm than fibrous astrocytes and form perivascular end-feet. The velate type of astrocyte contains lamellae (Fig. 27), arranged as single or multilamellar formations around elements of the neuropil. Gap junctions and puncta adhaerentia are formed between astrocytic membranes of astrocytic processes and lamellae (Bozhilova-Pastirova et al. 1989b,c).

We provide a description of the ultrastructural characteristics of the oligodendrocytes and contacts between astrocytic and oligodendrocytic membranes in apposition (Bozhilova-Pastirova et al. 1991a). Oligodendrocytes and their processes form astrocytic–oligodendrocytic gap junctions and oligodendrocytic–oligodendrocytic tight junctions.

From our freeze-etching replicas, it is apparent that the membrane of astrocytic processes and lamellae of the inferior olivary complex in the pigeon contains so-called assemblies (Fig. 28b), which are a characteristic feature of the mammalian astrocytic membranes (Dermietzel 1974; Landis and Reese 1974a; 1982; Hatton and Ellisman 1982; Anders and Brighman 1982; Massa and Mugnaini 1982; Gotow 1984; Waxman and Black 1984; Wujek and Reiner 1984; Mack and Wolburg 1986; Bozhilova-Pastirova et al. 1991b; Tao-Cheng et al. 1992). These orthogonal arrays of particles at the P-face and complementary pits at the E-face are more numerous on the membrane of the perivascular end-feet (Fig. 30) and less numerous in astrocytic membranes around the neuronal processes. The role of the assemblies is enigmatic (Dermitzel 1974). These structures would represent the morphological correlate of intramembranous enzyme complexes (Dermitzel 1974) or in an apposition to the blood vessels appear as the site of transport, but the nature of the transported substance and the direction of transport remain to be defined (Landis and Reese 1974a). There is indirect evidence that assemblies are composed of proteins and are connected with the cytoskeleton (Anders and Brightman 1982). According to Tao-Cheng et al. (1992), production of assemblies can be modulated by agents which raise cyclic AMP and activate protein kinase A but not protein kinase C.

Oligodendrocytic membranes of the cell body and their processes possess a heterogeneous population of intramembranous particles at both fractured faces as it is found in different regions of the central nervous system and cultured glial cells (Dermietzel et al. 1978; Massa and Mugnaini 1982, 1985; Waxman and Black 1984). In our freeze-etching replicas, heterogeneous oligodendrocyte-to-astrocyte gap junctions are found at the oligodendrocytic membrane of the cell body (Fig. 32) and the perinodal region (Fig. 33). These gap junctions comprised P-face particles having a diameter of 7.69±0.27 μm (SEM) with a range of 6–9 nm. Intramembranous particles of homogeneous astro-astrocytic gap junctions are quite uniform (9 nm in diameter) and are arranged in a hexagonal manner at the P-face, as it is found in glial cells of the cerebral cortex and in cultured glial cells (Massa and Mugnaini 1982, 1985). Oligodendrocyte-to-oligodendrocyte tight junctions are a linear specialization and are formed between oligodendrocytic cell bodies, cell body and process, and the outer tongue process and the turn of the myelin sheath membrane (Massa et al. 1984). Bands of diagonal arrows of particles and parallel particle-free strands on the P-face of the oligodendrocytic membrane of perinodal pockets are a constituent of the axonal-glial apposition in the perinodal region (Schnapp and Mugnaini 1975; Dermitzel et al. 1978; Waxman and Black 1984).

The gap junctions between glial cells would represent the morphological correlate of the electrotonic coupling or may serve to couple cells metabolically and to allow for intermembrane passage of substances (Bennett and Goodenough 1975; Massa and Mugnaini 1982; Waxman and Black 1984; Mugnaini 1986). In many mammalian tissues a 27 kDa integral membrane protein, the so-called connexon 32, as well as connexon 43 are identified as components of gap junctions (Dermitzel et al. 1984, 1989, 1991; Hertzberg and Spray 1985; Beyer et al. 1987; Nagi et al. 1988, 1989; Musil and Goodenough 1993). A hexametric arrangement of these proteins in apposition at the cell membrane forms an intercellular communicating channel (Bennett and Goodenough 1975; Paracchia 1980). Oligodendrocytic membranes form tight contacts across the extracellular space and resemble membrane interactions at intraperiod lines with myelin sheaths (Massa and Mugnaini 1985).

4.2.3
Neuronal Plasticity in the Inferior Olivary Complex

The specificity of olivary connections, neurochemical specificity of the functional properties of olivary neurons, and elements of the neuropil are one of the most important properties of the inferior olivary complex, but there are examples of plastic changes during development and in adult brains. Plasticity is an important property of the neurons (see Bloom 1985 for references) and historically this idea belongs to Ramón y Cajal (1906). According to Zilles (1992), the concepts of plasticity and specificity are not contradictory, but supplementary. Plastic changes in the human inferior olivary complex, known as olivary hypertrophy, have been shown to result from lesions of the cerebellum or the pontine tegmentum (see Ruigrok 1990b for references). In the cat, such a phenomenon can be induced by ablation of the cerebellum including the cerebellar nuclei (Verhaart and Voogd 1962; Boesten and Marani 1979; Boesten and Voogd 1985; Marani 1986). Plastic changes can be induced by injury but also occur spontaneously and are based on the same mechanisms (Calverley and Jones 1990).

The unusual elements, which have been observed in our material in the well-preserved olivary neuropil, may be accepted as plastic changes which can be explained in several ways. For example, intranuclear filamentous inclusions are observed in different parts of the brain in normal conditions (see Bozhilova-Pastirova and Ovtscharoff 1985 for references) but also in cases with autosomal dominant cerebellar ataxia with degeneration caused by CAG/glutamine repeat expansion in the SCA7 gene/protein

(Holberg et al. 1998), or in the cerebral cortex of aged rats (Bozhilova-Pastirova et al. 1990). In hibernating ground squirrels these filamentous inclusions are in combination with membrane-bound inclusions. Such membrane-bound inclusions are present in dendritic profiles and axon terminals. They are marked with reaction product for acidic phosphatase and may be an example of spontaneous degeneration.

The cytoplasmic inclusions, the so-called nematosomes, are also present in the inferior olivary complex of ground squirrels (Bozhilova-Pastirova and Ovtscharoff 1984), especially in females animals during the spring. These cytoplasmic inclusions are found in the developing inferior olivary complex (P1–P5) of the rat (Bourrat and Sotelo 1983), but not in adult rats. Nematosomes are not marked with reaction product for acidic phosphatase, but after 3-acetylpyridine-induced degeneration of the inferior olive are seen in the soma of Purkinje cells (Desclin and Colin 1980) or in other pathological cases (see Ockleford et al. 1987 for references). These plastic changes can be explained by the expression of the immediate-early gene *c-fos* and manipulations of the endocrine system (Lee et al. 1987; Sagar et al. 1988; Herrera and Robertson 1990b; Robertson 1992). Sex differentiation of the brain is also a manifestation of neuronal plasticity either under the control of sex steroids, or as a result of genomic regulation (Pilgrim and Hutchison; 1994). Parvalbumin is transiently upregulated during development and parvalbumin immunoreactivity in dorsal parts of the dorsal cap of the medial accessory olive, but it could not be demonstrated on paraffin sections of 14th-day postnatal rats (Wassef et al. 1992a,b). We observed parvalbumin-immunoreactive neurons in the rat inferior olivary complex at postnatal day 20 in the dorsal cap of the medial accessory olive. Females have greater numbers of parvalbumin-immunoreactive perikarya per mm2 than males. This sex difference is considered significant as it is found also in other parts of the brain (Ovtscharoff et al. 1992; Stefanova et al 1997a,b).

Submembranous dense bodies are observed in the normal inferior olivary neurons in the cat (Bozhilova and Ovtscharoff 1978; 1979). The results of other studies indicate that these bodies are present in the inferior olivary complex undergoing radical changes, e.g., in the hypertrophic inferior olive (De Zeeuw et al. 1990d). They may be one of the differences in reaction of the hypertrophic olivary neurons (Boesten and Marani 1979; Boesten and Voogd 1985; Marani 1986; Yagishita et al. 1986; Ruigrok et al. 1990b) or are particularly related to the expression of the immediate-early gene protein *c-fos* that is demonstrated in the specific olivary areas of rats walking on a rotating drum for 75 min (Ruigrok et al. 1996). Brain damage also may produce activation of *c-fos* (Dragunow and Robertson 1988) either near the site of damage or in parts remote from the damage (Dragunow and Robertson 1988; Herrera and Robertson 1990a,b; Robertson 1992).

Spheroids and different types of axonal alternations are present in the inferior olivary complex of healthy animals and are probably examples of spontaneous degeneration (Sotelo and Palay 1971; Chan-Palay 1977; Gavelová et al. 1983; Bozhilova-Pastirova 1990a) or a result of age-related changes (Bozhilova-Pastirova and Ichev 1990), but they are also found in the inferior olivary complex of neonate children with neuroaxonal dystrophy (Janota 1979). The responses to the injury or primary aging in the inferior olivary complex are olivary hypertrophy, loss of neurons, and activated astrocytes and are associated with positive reactions for synaptophysin, NF-2F11, NSA, chromographin, S-100β, GFAP (Gioccone et al. 1988; Bozhilova-Pastirova et al.

1989d; Grandi and Arcasi 1992; Ito et al. 1997), as well as with cases of progressive dysphasic dementia after cerebral atrophy (Kabayashi et al. 1990).

Astrocytic multilamellar capsules surround the altered axon terminals or myelinated axons in the inferior olivary complex (Bozhilova-Pastirova et al. 1991b) as in the other parts of the brain (Van Houten and Brower 1979). Reactive microglial cells swallow up multilamellar capsules containing degenerative neuronal material. According to Chan-Palay (1977), this process could be accepted as normal if the spontaneous degeneration is a part of a continuous process of repairing, which neurons normally undergo. Free postsynaptic densities are interpreted as synaptic contacts which have lost their presynaptic component and considered as an intermediate stage in a process of reinnervation (Westrum 1980; Spacek 1982; Tweedle and Hatton 1986; Bozhilova-Pastirova and Ichev 1990; Bozhilova-Pastirova 1990a). According to Buffo et al. (1998), axotomized inferior olivary neurons in the adult rat undergo regressive modifications leading to cell loss, but affected neurons are capable of expressing several markers related to regenerative processes, and at least a subset of them is resistant to the lesion.

Our observations on the intramembranous structure of the synaptic membrane as well as glial cell membranes show specific variations and probably reflect functional differences. In one case, plasticity is reflected by variations in the organization of the intramembranous particles at the E-face of the postsynaptic membrane, in the other case by differences in the number and distribution of the orthogonal particles.

5
Summary

In conclusion, the main results in this study are original or supportive to the debated assertions in the literature. The conclusions summarize the results in this monograph as follows:

5.1
Light Microscopy. Topography and Cell Types of the Inferior Olivary Complex

1. The present study demonstrates that the parcellation of the submammalian inferior olivary complex varies from small groups of cells in the frog, lizard, and tortoise to a column of cells in the carp and pigeon. The mammalian inferior olivary complex comprises three major subnuclei and several subgroups, but their relative size and position vary in the rat, cat, and human. In addition, we have provided the original description of the topography of the inferior olivary complex in the ground squirrel (*Citellus citellus L.*) and correlated our observations with data obtained from different mammalian species.
2. In submammalian (carp and pigeon) and mammalian (ground squirrel, cat, and human) species, variations between olivary neurons in the size and density, as described in this study, indicate a decrease in neuronal density from carp to human. It seems to be concomitant of an increase in mean neuronal area toward the human.

3. As shown in the present study, the pigeon inferior olivary complex has some characteristics in common with mammals. Single Golgi-stained neurons with ball-like appearance were rarely seen in the dorsal lamella, but often marginal reticular neurons were found to send their dendrites into both lamellae of the pigeon inferior olivary complex. These predominately large reticular neurons may take part in the local circuitry. But an obvious difference is the architecture of the dendritic tree. This, of course, reflects the size of the dendritic field area. The comparison of the measurements for packing densities of the olivary neurons per mm3 and its mean dendritic field area unequivocally shows that dendrites of the olivary neurons in the carp and pigeon are more overlapping, while olivary neurons in mammals have compact and restricted dendritic field.

4. Parvalbumin- and GABA-immunoreactive neurons were observed in the dorsal cap and small part of nucleus β on postnatal day 20 in male and female rats and are described in the present study. Moreover, there are sexual differences in the densities of the parvalbumin- and GABA-immunoreactive neurons of the dorsal cap and small part of nucleus β in 20-day-old rats. Females have a significantly greater number of immunoreactive neurons than males. These sexual differences might concern neuronal plasticity. Nonetheless, with respect to a classification of these olivary neurons in interneurons or projection neurons, no separation could be made.

5.2
Ultrastructure of the Neurons and Glial Cells
in the Inferior Olivary Complex

1. In general, the ultrastructure of the inferior olivary neurons as well as the reticular neurons in or around the lamellae of the pigeon resembles those in the mammalian species (rat, ground squirrel, and cat). One of the most extraordinary morphological features in the inferior olivary complex is extensive membrane apposition, which occurs between somata, dendrites, soma, and dendrites. We provide original evidence that some neurons, predominantly in the medial accessory olive, possess cilia with a modified pattern.

2. Dendritic elements in the neuropil of the pigeon olivary complex were found to form close dendro-dendritic appositions characterized by puncta adhaerentia in the central core of the glomerular-like synaptic complexes. However, the evidence for dendro-dendritic gap junctions between pigeon olivary neurons was not established in this study. On the basis of these data it might be hypothesized that electrotonic transmission occurs later in phylogeny.

3. The present study extends the observations on the synaptic organization in the submammalian and mammalian inferior olivary complex by analysis of the intramembranous structure of the postsynaptic membrane, presynaptic membrane, gap junctions, and puncta adhaerentia. The intramembranous structure of dendro-dendritic gaps junction (only present in mammals) and puncta adhaerentia particle aggregates is reported for the first time in the inferior olivary complex. A correlation between the fine structure of the synaptic contact zone in thin sections and the intramembranous structure of the presynaptic and postsynaptic membranes in the freeze-etching replicas was made. On the other hand, we correlated

the intramembranous structure of the postsynaptic E-face with the type of transmitter and its action. The size and packing density of the intramembranous particles at both faces of presynaptic and postsynaptic membranes were measured and compared.

4. Morphological characteristics of the glial cells were studied in Golgi-stained preparations and correlated with electron-microscopical data from thin sections and freeze-etching replicas. In Golgi-stained preparations of the inferior olivary complex (pigeon, rat, ground squirrel, and cat), fibrous, protoplasmic, and velate astrocytes were described. The velate astrocytes were predominantly of the glial cell type in the olivary neuropil. Slightly larger protoplasmic astrocytes could be seen in the neuropil, as well as in the surrounding white matter, accompanied by fibrous astrocytes and oligodendrocytes, too. The normal ultrastructure of the glial cells in the pigeon, rat, ground squirrel, and cat inferior olivary complex was studied by using thin sections and freeze-etching replicas. In the freeze-etching replicas, astrocytic processes were unequivocally identified by orthogonal arrays of small intramembranous particles or assemblies associated with the P-face and complementary pits at the E-face. From our freeze-etching study, it is apparent that the membrane of astrocytic processes and lamellae of the inferior olivary complex in the pigeon contains assemblies, which are a characteristic feature of the mammalian astrocytic membranes. Numerous elongated intramembranous particles, fewer globular intramembranous particles, and the presence of tight junctions between adjacent oligodendrocytic membranes characterized oligodendrocytic membrane. Interastrocytic and astrocyte-to-oligodendrocytic gap junctions were also observed. So, it would appear that both at the light- and electron-microscopical level, glial cells in the pigeon inferior olivary complex resemble glial cells in the mammals (ground squirrel, rat, and cat).

5. In general, astrocytes display the usual cytological features in the normal inferior olivary neuropil, however, some of these glial cells appear to be reactive. Such astrocytes contain numerous gliofilaments and are GFAP-immunoreactive. Their processes surround degenerating neuronal profiles within multilamellar capsules. Some of the encapsulated degenerating debris appears engulfed by microglial cells. Other signs of reactivity include the presence of numerous assemblies on the P-face and complementary pits on the E-face of the freeze-etching membranes in the astrocytic processes containing numerous gliofilaments and the astrocytic lamellae within the multilamellar capsules.

6. In this study we present examples of unusual changes in the neuronal fine structure such as: intranuclear inclusions, spheroids, and other types of axonal alternations, which are marked with reaction product for acid phosphatase in the inferior olivary complex of normal animals. In the ground squirrel inferior olivary complex we found nematosomes predominately in females during the spring. These results are an important morphological expression of the adaptive property of the olivary neurons.

Acknowledgements. We are greatly indebted to Prof. E. Marani who suggested the idea of writing this monograph and supported us with his valuable advise. The computer-assisted measures were carried out in collaboration with Mrs. D. Brazitsova. The authors would like to thank Assoc. Prof. R. Dimova†, Dr. L. Surchev, and Dr. T. Christova for expert technical assistance. We greatly appreciate critical reading of the manuscript by Dr. N. Stefanova and Prof. K. Usunoff.

References

Abercrombie M (1946) Estimation of nuclear populations from microtome sections. Anat Rec 94: 239–247

Akert K, Pfenninger K, Sandri C, Moor H (1972) Freeze-etching and cytochemistry of vesicles and membrane complexes in synapses of central nervous system. In: GD Pappas and DP Purpura (eds) Structure and function of synapses. New York. Raven Press

Albus JS (1971) A theory of cerebellar function. Math Biosci 10: 25–61

Alonso A, Blanco NJ, Paino CL, Rubia FJ (1986) Distribution of neurons in the main cuneate nucleus projecting to the inferior olive in the cat. Evidence that they differ from those directly projecting to the cerebellum. Neuroscience 18: 67196683

Anders JJ, Brighman MW (1982) Particle assemblies in astrocytic plasma membranes are rearranged by various agents *in vitro* and cold injury *in vivo*. J Neurocytol 11: 1009–1029

Angaut P, Alvarado-Mallart RM, Sotelo C (1985) Compensatory climbing fiber innervation after unilateral pedunculotomy in the newborn rat: Origin and topographic organization. J Comp Neurol 236: 161–178

Angaut P, Cicirata F (1982) Cerebello-olivary projections in the rat: An autoradiographic study. Brain Behav Evol 1: 43–57

Angaut P, Sotelo C (1987) The dento-olivary projection in the rat as a presumptive GABAergic link in the olivo-cerebelo-olivary loop. An ultrastructural study. Neurosci Lett 83: 227–231

Angaut P, Sotelo C (1989) Synaptology of cerebello-olivary pathway. Double labelling with anterograde axonal tracing and GABA immunocytochemistry in the rat. Brain Res 479: 361–365

Aoki E, Semba R, Kato K, Kashiwamata S (1987) Purification of specific antibody against aspartate and immunocytochemical localization of aspartergic neurons in the rat brain. Neuroscience 21: 755–765

Apps R (1990) Columnar organization of the inferior olive projection to the posterior lobe of the rat cerebellum. J Comp Neurol 302: 236–254

Apps R (1998) Input-output connections of the 'hindlimb' region of the inferior olive in cats. J Comp Neurol 399: 513–529

Apps R (1999) Movement-related gating of climbing input to cerebella cortical zones. Prog Neurobiol 57: 537–562

Apps R, Lee S (1999) Gating of transmission in climbing fibre paths to cerebellar cortical C1 and C3 zones in the rostral paramedian lobule during locomotion in the cat. J Physiol (Lond) 516 (Pt3): 875–883

Arends JA, Voogd J (1989) Topographical aspects of the olivocerebellar system in the pigeon. In: P Strata (ed) The Olivocerebellar system in motor control. Exp Brain Res series 17, Springer-Verlag, Berlin, pp 52–57

Ariens Kappers CU, Huber GC, Grosby EC (1936) The comparative anatomy of nervous system of vertebrates, including man. In: Macmillan Co, New York, pp. 668–689

Armengol JA, Lopes-Raman A (1996) Pattern of degeneration of the rat inferior olivary complex after the early postnatal axotomy of the olivocerebellar projection. Histol Histopathol 11: 379–388

Armstrong DM (1974) Functional significance of connections of the inferior olive. Physiol Rev 54: 358–417

Armstrong DM, Harvey RF, Schild RF (1974) Topographical localization in the olivo-cerebellar projection: An electrophysiological study in the cat. J Comp Neurol 154: 287–302

Armstrong DM, Schild RF (1980) Location of the spinal cord of neurons projecting directly to the inferior olive in the cat. In: Courville J, de Montigny C, Lamare Y (eds) The inferior olivary nucleus. Anatomy and physiology. Raven Press, New York 1980, pp 125–144

Armstrong RC, Clarke PG (1979) Neuronal death and the development of the pontine nuclei and inferior olive in the chick. Neuroscience 4: 1635–1649

Azizi SA, Woodward DJ (1987) Inferior olivary nuclear complex of the rat: Morphology and comments on the principles of organization within the olivocerebellar system. J Comp Neurol 263: 467–484

Balaban CD, Beryozkin G (1994) Organization of vestibular nucleus projections to the caudal dorsal cap of Kooy in rabbits. Neuroscience 62: 1217–1236

Bangma GC, ten Donkelaar HJ (1982) Afferent connections of the cerebellum in various types of reptiles. J Comp Neurol 207: 255–273

Barmack NH (1996) GABAergic pathways convey vestibular information to beta nucleus and dorsomedial cell column of the inferior olive. Ann N Y Acad Sci 781: 541–552

Barmack NH, Fredette BJ, Mugnaini E (1998) Parasolitary nucleus: a source of GABAergic vestibular information to the inferior olive of rat and rabbit. J Comp Neurol 392: 352–372

Benardo LS, Foster RE (1986) Oscillatory behavior in the inferior olive neurons: mechanism, modulation, cell aggregates. Brain Res Bull 17: 773–784

Bennett MVL, Goodenough DA (1975) Gap junctions, electrotonic coupling and intracellular communication. Neurosci Res Prog Bull 16: 373–486

Berkley KJ, Worden IG (1978) Projections to the inferior olive of the cat. I. Comparisons of input from the dorsal column nuclei, cervical nucleus, spino-olivary pathways, the cerebral cortex and cerebellum. J Comp Neurol 180: 237–252

Bernard JF (1987) Topographical organization of olivocerebellar and corticonuclear connections in the rat. An WGA-HRP study. I. Lobules IX, X and the flocculus. J Comp Neurol 263: 241–258

Beyer EE, Paul DL, Goodenough DA (1987) Connexin 43: A protein from rat homologous to a gap junction protein from liver. J Cell Biol 105: 2621–2929

Bishop GA (1984) The origin of the reticulo-olivary projection in the rat: a retrograde horseradish peroxidase study. Neuroscience 11: 487–496

Bishop GA, Ho RH (1986) Cell bodies of origin of serotonin-immunoreactive afferents to the inferior olivary complex of the rat. Brain Res 399: 369–373

Bishop GA, King JS (1986) Reticulo-olivary circuits: An intracellular HRP study in the rat. Brain Res 371: 133–145

Bishop GA, Ho RH (1986) Cell bodies of origin of serotonin-immunoreactive afferents to the inferior olivary complex of the rat. Brain Res 399: 369–373

Bloom FE (1985) CNS plasticity: A survey of opportunities. In: Bignami A, Bloom FE, Bolis CL, Adeloy A (eds) Central nervous system plasticity and repair. Raven Press, New York, pp 3–11

Boesten AJP, Marani E (1979) A light- and electron-microscopical study of olivary hypertrophy in the cat. Neurosci Lett 8: 158

Boesten AJP, Voogd J (1975) Projections of the dorsal column nuclei and the spinal cord on the inferior olive in the cat. J Comp Neurol 161: 215–238

Boesten AJP, Voogd J (1985) Hypertrophy of neurons in the inferior olive after cerebellar ablations in the cat. Neurosci Lett 61: 49–54

Bourrat F, Sotelo C (1983) Postnatal development of the inferior olivary complex in the rat. I. An electron microscopic study of the medial accessory olive. Dev Brain Res 8: 291–310

Bourrat F, Sotelo C (1988) Migratory pathways and neuritic differentiation of olivary neurons in the rat embryo. Axonal tracing study using the in vitro slab technique. Dev Brain Res 39:19–37

Bourrat F, Sotelo C (1990a) Early development of the rat precerebellar system: migratory routes, selective aggregation and neuritic differentiation of the inferior olive and lateral nucleus neurons. A overview. Arch Ital Biol 128: 151–170

Bourrat F, Sotelo C (1990b) Migratory pathways and selective aggregation of the lateral reticular neurons in rat embryo: a horseradish peroxidase in vitro study, with special reference to migration patterns of the precerebellar nuclei. J Comp Neurol 294: 1–13

Bourrat F, Sotelo C (1991) Relationships between neuronal birthdates and cytoarchitecture in the rat inferior olivary complex. J Comp Neurol 313: 509–521

Bowman JP, Sladek JR Jr. (1973) Morphology of the inferior olivary complex of the rhesus monkey (Macaca mulatta). J Comp Neurol 152: 299–316

Bowman MH, King J (1973) The conformation, cytology and synaptology of the opossum inferior olivary nucleus. J Comp Neurol 148: 491–524

Bozhilova A (1976) Structure of the inferior olivary nucleus. I. Reptile (in Bulgarian). Med-Biol Problems 4: 61–65

Bozhilova A, Ovtscharoff W (1979) Synaptic organization of the medial accessory olivary nucleus of the cat. J Hirnforsch 20: 19–28

Bozhilova AI, Ovtscharoff WA (1978) Submembraneous dense bodies in the medial accessory olivary nucleus of the cat. Compt rendus de l'Acad bulg Sci 31: 755–757

Bozhilova -Pastirova A (1990a) Ultrastructure of the inferior olivary complex in the ground squirrel (Citellus Citellus L.). Dendrites. Age-related changes (in Bulgarian). Exsperim med i morf 29-3: 9–17

Bozhilova-Pastirova A (1990b) Ultrastructure of the inferior olivary complex in the ground squirrel (*Citellus Citellus L.*) (in Bulgarian). Exsperim med i morf 29-3: 1–8

Bozhilova-Pastirova A, Ichev K (1989) Free postsynaptic densities in the inferior olivary complex of the ground squirrel (*Citellus Citellus L.*) (in Bulgarian). Exsperim med i morf 29-1: 8–14

Bozhilova-Pastirova A, Ovtscharoff W (1985) Nuclear inclusions in the inferior olivary nucleus of the cat and ground squirrel (*Citellus citellus L.*). Z Microsk Anat Forsch, Leipzig 99: 725–734

Bozhilova-Pastirova A, Ovtscharoff W (1995a) The morphology of inferior olivary complex in phylogeny. Eur J Morphol 33: 176

Bozhilova-Pastirova A (1990c) Structural and ultrastructural study of the inferior olivary complex in the ground squirrel (*Citellus citellus L.*) (in Bulgarian). Exsperim med i morf 29-4: 1–7

Bozhilova-Pastirova A (1998) Freeze-etching study of the axosomatic synapses in the rat sensorimotor cortex. Eur J Morphol 36: 189–200

Bozhilova-Pastirova A, Ichev K (1990) Ultrastructure of the inferior olivary complex of the ground squirrel (*Citellus citellus L.*). I. Neuropil. Age-related changes in axons (in Bulgarian). Exsperim med i morf 30-4: 8–14

Bozhilova-Pastirova A, Ichev K, Ovtscharoff W (1989a) Light microscopical and ultrastructural examination of the inferior olivary complex of the pigeon (*Columba livia*) (in Bulgarian). Exsperim med i morf 28-2: 9–14

Bozhilova-Pastirova A, Ichev K, Ovtscharoff W (1989b) Morphology of glial cells in the inferior olivary complex of the pigeon (*Columba livia*). I. Astrocytes (in Bulgarian). Exsperim med i morf 28-8: 22–26

Bozhilova-Pastirova A, Ichev K, Ovtscharoff W (1989c) Morphology of glial cells in the inferior olivary complex of the pigeon (*Columba livia*). I. Oligodendrocytes (in Bulgarian). Exsperim med i morf 28-3: 26 31

Bozhilova-Pastirova A, Ichev K, Ovtscharoff W (1989d) Distribution of the glial fibrillary acidic protein in the inferior olivary complex of the cat, ground squirrel and pigeon. In: Histochemical methods and their application in the science and diagnosis. Gegenbaurs morphol Jahrb Leipzig 135 2 S, 310

Bozhilova-Pastirova A, Ichev K, Ovtscharoff W (1991a) Morphology of glial cells in the inferior olivary complex of the ground squirrel (*Citellus citellus L.*). I. Oligodendrocytes (in Bulgarian). Exsperim med i morf 30-1: 15–21

Bozhilova-Pastirova A, Ichev K, Ovtscharoff W, Donchev V (1990) Ultrastructural study of the sensorimotor cortex in the aged rats. I. Intranuclear inclusions in neurons and astrocytes. (in Bulgarian). Exsperim med i morf 29-1: 1–7

Bozhilova-Pastirova A, Itchev K, Christova T (1991b) Morphology of the astrocytes in the inferior olivary complex of the ground squirrel (*Citellus citellus L.*). Verh Anat Ges 85. Anat Anz Suppl 170: 685–686

Bozhilova-Pastirova A, Ovtschariff W (1984) Nematosomes in neurons of the inferior olivary nucleus in the ground squirrel (*Citellus citellus L.*) (in Bulgarian). In: Contemporary problems of Neuromorphology. v 13–14, Bulgarian Academy of Sciences, Sofia, pp 161–165

Bozhilova-Pastirova A, Ovtscharoff W (1995b) Structure of the synaptic junctions in the rat sensorimotor cortex: freeze-etching study of neuronal gap junctions. Neurosci Lett 201: 265–267

Bozhilova-Pastirova A, Ovtscharoff W (1996a) Intramembranous structure of the postsynaptic membrane in the rat sensorimotor cortex. Neurosci Lett 206: 29–132

Bozhilova-Pastirova A, Ovtscharoff W (1996b) Structure of the synaptic junctions in the rat sensorimotor cortex. Freeze-etching study of axodendritic synapses. Eur J Morphol 34: 363–373

Bozhilova-Pastirova A, Ovtscharoff W (1999) Intramembranous structure of synaptic membranes with special reference to spinules in the rat sensorimotor cortex. Eur J Neurosci 11: 1843–1846

Braak H (1970) Über die Kerngebiete des menschlichen Hirnstammes. I. Oliva inferior, Nucleus conterminalis und Nucleus vermiformis corpuris restiformis. Z Zellforsch 105: 442–456

Breazile JE (1967) The cytoarchitecture of the brain stem of the domestic pig. J Comp Neurol 129: 169–188

Brodal A (1940) Experimentalle Untersuchungen über die olivocerebellare Lokalisation. Z Gesamte Neurol Psychiatr 169: 1–153

Brodal A (1980) Olivocerebellocortical projection in the cat as demonstrated with the method of retrograde transport of horseradish peroxidase. 2. Topographical pattern in relation to the longitudinal subdivision of the cerebellum. In: Courville J, de Montigny C, Lamare Y (eds). The inferior olivary nucleus. Anatomy and physiology. Raven Press, New York, pp 87–205

Brodal A, Kawamura K (1980) Olivocerebellar projections: A review. Adv Anat Embryol Cell Biol 64: 1–140

Brodal A, Torvik A (1957) Uber den Ursprung der sekundaeren vestibulo-cerebellaren Fasern bei der Katze. Arch Psychiat Nervenkr 195: 550–567

Brodal P, Brodal A (1981) The olivocerebellar projection in the monkey. Experimental studies with the method of retrograde tracing of horseradish peroxidase. J Comp Neurol 201: 375–393

Brodal P, Brodal A (1982) Further observations on the olivocerebellar projection in the monkey. Exp Brain Res 45: 71–83

Brown JT, Chan-Palay V, Palay SL (1977) A study of afferent input to the inferior olivary complex in the rat by retrograde axonal transport of horseradish peroxidase. J Comp Neurol 176: 1–22

Buffo A, Fronte M, Oestreicher AB, Rossi F (1998) Degenerative phenomena and reactive modifications of the adult rat inferior olivary neurons following axotomy and disconnection from their targets. Neuroscience 85: 587–604

Buisseret-Delmas C (1980) An HRP study of the afferents to the inferior olive in cat. I-cervical spinal and dorsal column nuclei projections. Arch Ital Biol 118: 270–286

Buisseret-Delmas C (1988a) Sagittal organization of the olivocerebellonuclear pathway in the rat. I. Connections with the nucleus fastigii and the nucleus vestibularis lateralis. Neurosci Res 5: 475–493

Buisseret-Delmas C (1988b) Sagittal organization of the olivocerebellonuclear pathway in the rat. I. Connections with the nucleus interpositus. Neurosci Res 5: 494–512

Buisseret-Delmas C, Angaut P (1993) The cerebellar olivo-corticonuclear connections in the rat. Prog Neurobiol 40: 63–87

Buisseret-Delmas C, Bantini C, Compoint C, Daniel H, Menetrey D (1989) The GABAergic neurones of the cerebellar nuclei: Projection to the caudal inferior olive and to the bulbar reticular formation. In: P Strata (ed) The Olivocerebellar system in motor control. Exp Brain Res Series 17, Springer-Verlag, Berlin, pp 108–110

Buisseret-Delmas C, Batini C (1978) Topology of the pathways to the inferior olive: an HRP study in cat. Neurosci Lett 10: 207–215

Bull MS, Mitchell SK, Berkley KJ (1990) Convergent inputs to the inferior olive from the dorsal column nuclei and pretectum in the cat. Brain Res 525: 1–10

Butter-Ennever JA, Cohen B, Horn AK, Reisine H (1996) Efferent pathways of the nucleus of the optic tract in monkey and their role in eye movements. J Comp Neurol 373: 90–107

Calverley RKS, Jones DG (1990) Contribution of dendritic spines and perforated synapses to synaptic plasticity. Brain Res Rev 15: 215–249

Campbell NC, Armstrong DM (1983) Topographical localization in the olivocerebellar projection in the rat. An autoradiographic study. Brain Res 275: 235–249

Chan-Palay V (1977) Cerebellar dentate nucleus. Organization, cytology and transmitters. Springer Verlag, Berlin Heidelberg, New York

Chan-Palay V, Palay SL (1972) The form of velate astrocytes in the cerebellar cortex of monkey and rat: High voltage electron microscopy of rapid Golgi preparations. Z Anat Entwickl-Gesch 138: 1–19

Chan-Palay V, Palay SL, Brown JT, van Itallie C (1977) Sagittal organization of olivocerebellar and reticulocerebellar projections: autoradiographic studies with 35S-methionine. Exp Brain Res 30: 561–576

Chedotal A, Sotelo C (1993) The 'creeper stage' in cerebellar climbing fiber synaptogenesis precedes the 'pericellular nest'-ultrastructural evidence with parvalbumin immunocytochemistry. Dev Brain Res 76: 207–220

Clarke PGH (1977) Some visual and other connections to the cerebellum of the pigeon. J Comp Neurol 174: 535–552

Cochran SL, Hackett JT (1977) The climbing fiber afferent system of the frog. Brain Res 121: 362–367

Cook JR, Wiesendanger M (1976) Input from trigeminal cutaneus afferents to neurons of the inferior olive in rats. Exp Brain Res 26: 193–202

Courville J, Faraco-Cantini F (1980) Topography of olivocerebellar projection: An experimental study in cat with an autoradiographic tracing method. In: Courville J, de Montigny C, Lamare Y (eds) The inferior olivary nucleus. Anatomy and physiology. Raven Press, New York, pp 235–277

Courville J, Faraco-Cantini F, Legendre A (1983a) Detailed organization of cerebello-olivary projections in the cat. An autoradiographic study. Arch Ital Biol 121: 219–236

Courville J, Faraco-Cantini F, Marson L (1983b) Projections from the reticular formation of the medulla, the spinal trigeminal and lateral reticular nuclei to the inferior olive. Neuroscience 9: 129–139

Cunningham JJ, Sherrard RM, Bedi KS, Renshaw GM, Bower AJ (1999) Changes in the numbers of neurons and astrocytes during the postnatal development of the rat inferior olive. J Comp Neurol 406: 375–383

Dahl HA (1963) Fine structure in the rat cerebral cortex. Z Zellforsch 60: 369–386

De Zeeuw CI, Gerris NM, Voogd J, Leonard CS, Simpson JI (1994) The rostral dorsal cap and ventrolateral outgrowth of the rabbit inferior olive receive a GABAergic input from dorsal group y and ventral dentate nucleus. J Comp Neurol 341: 42096432

De Zeeuw CI, Hertzberg EL, Mugnaini E (1995) The dendritic lamellar body: a new neuronal organelle putatively associated with dendrodendritic gap junctions. J Neurosci 15: 1587–1604

De Zeeuw CI, Holstege JC, Calkoen F, Ruigrok TJH, Voogd J (1988) A new combination of WGA-HRP anterograde tracing and GABA immunocytochemistry applied to afferents of the cat inferior olive at the ultrastructural level. Brain Res 447: 369–375

De Zeeuw CI, Holstege JC, Ruigrok TJH, Voogd J (1989a) The cerebellar, mesodiencephalic and GABAergic innervation of the glomeruli in the cat inferior olive. A comparison at the ultrastructural level. Exp Brain Res 17: 111–116

De Zeeuw CI, Holstege JC, Ruigrok TJH, Voogd J (1989b) Ultrastructural study of the GABAergic, cerebellar and mesodiencephalic innervation of the cat medial accessory olive; anterograde tracing combined with immunocytochemistry. J Comp Neurol 284: 12–35

De Zeeuw CI, Holstege JC, Ruigrok TJH, Voogd J (1990a) Mesodiencephalic and cerebellar terminals terminate upon the same dendritic spines in glomeruli of the cat and rat inferior olive: An ultrastructural study using a combination of [3H]-leucin and wheat germ agglutinin coupled horseradish peroxidase anterograde tracing. Neuroscience 34: 645–655

De Zeeuw CI, Hoogenraat CC, Goedknegt E, Hertzberg E, Neubauer A, Grosveld F, Galjart N (1997a) CLIP-115, a novel brain specific cytoplasmic linker protein, mediates the localization of dendritic lamellar bodies. Neuron 19: 1187–1199

De Zeeuw CI, Koekkoek SKE (1997) Signal processing in the C2 module of the flocculus and its role in head movement control. In: De Zeeuw CI, Strata P, Voogd J (eds) The cerebellum: from structure to control. Elsevier, Amsterdam, Lausanne, New York, Oxford, Shannon, Tokyo. Pogr Brain Res 114, 299–320

De Zeeuw CI, Lang EJ, Sugihara I, Ruigrok TJH, Eisenman LM, Mugnaini E, Llinás R (1996) Morphological correlates of bilateral synchrony in the rat cerebellar cortex. J Neurosci 16: 3412–3426

De Zeeuw CI, Ruigrok TJH, Holstege JC, Jansen HG, Voogd J (1990b) Intracellular labelling of neurons in the medial accessory olive of the cat: II. Ultrastructure of dendritic spines and their and their GABAergic innervation. J Comp Neurol 300: 478–494

De Zeeuw CI, Ruigrok TJH, Holstege JC, Schalekamp MPA, Voogd J (1990c) Intracellular labeling of neurons in the medial accessory olive of the cat: III. Ultrastructure of axon hillock and initial segment and their GABAergic innervation. J Comp Neurol 300: 495–510

De Zeeuw CI, Ruigrok TJH, Schalekamp MPA, Boesten AJP, Voogd J (1990d) Ultrastructural study of the cat hypertrophic inferior olive following anterograde tracing, immunocytochemistry and intracellular labeling. Eur J Morphol 28: 240–255

De Zeeuw CI, Simpson JI, Hoogenraad CC, Galjart N, Koekkoek SK, Ruigrok TJ (1998) Microcircuitry and function of the inferior olive. Trends Neurosci 21: 391–400

De Zeeuw CI, Van Alphen AM, Hawkins RK, Ruigrok TJH (1997b) Climbing fibre collaterals contact neurons in the cerebellar nuclei that provide a GABAergic feedback to the inferior olive. Neuroscience 80: 981–986

De Zeeuw CI, Wentzel P, Mugnaini E (1993) Fine structure of dorsal cap of the inferior olive and its GABAergic and non-GABAergic input from the nucleus prepositus hypoglossi in rat and rabbit. J Comp Neurol 327: 63–82

Del Cerro MP, Snider RS (1967) Cilia in the cerebellum of immature and adult rats. J Microscopie 6: 515–518

Dermietzel R (1974) Junctions in the central nervous system of the cat. III. Gap junctions and membrane-associated orthogonal particle complexes (MOPC) in the astrocytic membrane. Cell Tiss Res 149: 121–135

Dermietzel R, Hertzberg EL, Kesslar JA, Spray DC (1991) Gap junctions between cultured astrocytes: immunocytochemical, molecular, electrophysiological analysis. J Neurosci 11: 1421–1432

Dermietzel R, Leibstein A, Frixen U, Janssen-Timmen U, Teaub O, Willecke K (1984) Gap junctions in several tissues share antigenic determinants with liver gap junction. J Embo 3: 2261–2270

Dermietzel R, Schünke D, Leibstein A (1978) The oligodendrocytic junctional complex. Cell Tiss Res 193: 61–72

Dermietzel R, Traub O, Hwang TK, Beyer E, Bennett MVL, Spray DC, Willecke K (1989) Differential expression of three gap junction proteins in developing and mature brain tissues. Proc Natl Acad Sci USA 86: 10148–10152

Desclin JC (1974) Histological evidence supporting the inferior olive as the major source of cerebellar climbing fibers in the rat. Brain Res 77: 365–384

Desclin JC, Colin F (1980) The olivocerebellar system. II. Some ultrastructural correlations of inferior olive destruction in the rat. Brain Res. 187: 29–46

Dietrichs E, Walberg F (1985) The cerebellar nucleo-olivary and olivo-cerebellar nuclear projections in cat as studied with anterograde and retrograde transport in the same animal after implantations of crystalline WGA-HRP. II. The fastigial nucleus. Anat Embryol 173: 253–261

Dietrichs E, Walberg F (1986) The cerebellar nucleo-olivary and olivo-cerebellar nuclear projections in cat as studied with anterograde and retrograde transport in the same animal after implantations of crystalline WGA-HRP. III. The interposed nucleus. Brain Res 373: 373–383

Dietrichs E, Walberg F (1989) Direct bidirectional connections between the inferior olive and the cerebellar nuclei. In: Strata P (ed) The olivocerebellar system in motor control. Exp Brain Res Series 17. Berlin. Springer-Verlag, pp 61–81

Dietrichs E, Walberg F, Nordby T (1985) The cerebellar nucleo-olivary and olivo-cerebellar nuclear projections in cat as studied with anterograde and retrograde transport in the same animal after implantations of crystalline WGA-HRP. I. The dentate nucleus. Neurosci Res 3: 52–70

Dom R, King JS, Martin GF (1973) Evidence for two direct cerebello-olivary connections. Brain Res 57: 498–501

Dragunow M, Robertson HA (1988) Brain injury induces c-fos protein(s) in nerve and glial-like cells in adult mammalian brain. Brain Res 295–299

Eccles JC, Llinás R, Sasaki K (1966) The synaptic excitatory synaptic action of climbing fibers on the Purkinje cells of the cerebellum. J Physiol London 182: 268–296

Eisenman LM (1981) Olivocerebellar projections to the pyramis and cupula pyramis in the rat: differential projections to parasagittal zones. J Comp Neurol 199: 65–76

Escobar A, Sampedro ED, Dow RS (1968) Quantitative data on the inferior olivary nucleus in man, cat and vampire bat. J Comp Neurol 132: 397–404

Fairén A. Peters A, Saldanha J (1977) A new procedure for examining Golgi impregnated neurons by light and electron microscopy. J Neurocytol 6: 311–337

Feirabentd HKP (1990) Development of longitudinal patterns in the cerebellum of the chicken (*Gallus domesticus*): A cytoarchitectural study on the genesis of cerebellar modules. Eur J Morphol 28: 169–223

Finger TE (1978) Cerebellar afferents in teleost catfish (*Ictaluridae*). J Comp Neurol 181: 173–181

Foster GA, Roberts PJ (1983) Neurochemical and pharmacological correlates of inferior olive destruction in the rat: Attenuation of events mediated by an endogenous glutamate-like substance. Neuroscience 8: 277–284

Foster RE, Peterson BE (1986) The inferior olivary complex of the guinea pig: Cytoarchitecture and cellular morphology. Brain Res Bull 17: 785–800

Fredette BJ, Adams JC, Mugnaini E (1992) GABAergic neurons in the mammalian inferior olive and ventral medulla detected by glutamate decarboxylase immunocytochemistry. J Comp Neurol 321: 501–514

Fredette BJ, Mugnaini E (1991) The GABAergic cerebello-olivary projection in the rat. Anat Embryol 184: 225–243

Freedman SL, Voogd J, Vielvoye GJ (1977) Experimental evidence for climbing fibers in the avian cerebellum. J Comp Neurol 175: 143–252

Freund TF (1989) GABAergic septohippocampal neurons contain parvalbumin. Brain Res 478: 375–381

Fujita M (1982) Adaptive filter model of the cerebellum. Biol Cybern 45: 195–206

Furber SE (1983) The organization of the olivocerebellar projection in the chicken. Brain Behav Evol 22: 198–211

Furber SE (1984) A Golgi study of the development of the inferior olivary nuclear complex in the chicken. J Comp Neurol 225: 244–258

Gavelová M, Badonic T, Mitro A (1983) Altered axons and axon terminals in the nucleus gracilis of the dog. J Hirnforsch 24: 399–404

Gellman R, Houk JC, Gibson AR (1983) Somatosensory properties of the inferior olive of the cat. J Comp Neurol 215: 228–243

Gerrits NM, Voogd J, Magras N (1985) Vestibular afferents of the inferior olive and the vestibulo-olivo-cerebellar climbing fiber pathway to the flocculus in the cat. Brain Res 332: 325–336

Gioccone G, Tagliavini F, Street JS, Ghetti B, Bugiani O (1988) Progressive supranuclear palsy with hypertrophy of the olives. An immunocytochemical study of the cytoskeleton of argyrophilic neurons. Acta Neuropathol 77: 14–20

Gomori G (1952) Microscopical histochemistry. Principles and practice. University of Chicago Press. Chicago, pp189–194

Gonzalez A, ten Donkelaar HJ, de Boer-van Huizen R (1984) Cerebellar connections in *Xenopus leavis*. An HRP study. Anat Embryol 169: 167–176

Gotow T (1984) Cytochemical characteristics of the plasma membranes specialized with numerous orthogonal arrays. J Neurocytol 13: 431–448

Gotow T, Sotelo C (1987) Postnatal development of the inferior olivary complex in the rat. IV. Synaptogenesis of GABAergic afferents, analyzed by glutamic acid decarboxylase immunocytochemistry. J Comp Neurol 263: 526–552

Grandi D, Arcari ML (1992) Neuronal aspects and plasticity of inferior olivary complex and nucleus dentatus. Acta Biomed Ateneo Parmense 63: 17–25

Graybiel AM, Nauta HJW, Lasek RJ, Nauta WJH (1973) A cerebello-olivary pathway in the cat: An experimental study using autoradiographic tracing techniques. Brain Res 58: 205–211

Gregg KV, Bishop GA (1997) Peptide localization in the mouse inferior olive. J Chem Neuroanat 12: 211–220

Grill WE (1970) Unitary multiple-spiked responses in the cat inferior olive nucleus. J Neurophysiol 33: 199–209

Groenewegen HJ, Boesten AJP, Voogd J (1975) The dorsal column nuclear projection to the nucleus ventralis posterior lateralis thalami and the inferior olive in the cat: an autoradiographic study. J Comp Neurol 162: 505–518

Groenewegen HJ, Voogd J (1977) The parasagittal zonation within the olivocerebellar projection. I. Climbing fibre distribution in the vermis of cat cerebellum. J Comp Neurol 174: 417–488

Groenewegen HJ, Voogd J, Freemen SL (1979) The parasagittal zonation within the olivocerebellar projection. II. Climbing fibre distribution in the intermediate and hemispheric parts of cat cerebellum. J Comp Neurol 183: 551–602

Grover BG, Grüsser-Cornehls U (1984) Cerebellar afferents in the frogs *Rana esculenta* and *Rana temporaria*. Cell Tiss Res 237: 259–267

Gwyn DG, Nicholson GP, Flumerfelt BA (1977) The inferior olivary nucleus of the rat: A light and electron microscope study. J Comp Neurol 174: 489–520

Harris KM, Landis DMD (1986) Membrane structure at synaptic junctions in area CA1 of the rat hippocampus. Neuroscience 19: 857–872

Hatton JD, Ellisman MH (1982) The distribution of orthogonal arrays in the freeze-fractured rat medial eminence. J Neurocytol 11: 335–349

Heckroth JA, Goldowitz D, Eisenman (1990) Olivocerebellar fiber maturation in normal and lurcher mutant mice: defective development in lurcher. J Comp Neurol 291: 415–430

Herrera DG, Robertson HA (1990a) Application of potassium chloride to the brain surface induces the c-fos proto-oncogene. Brain Res 510: 166–170

Herrera DG, Robertson HA (1990b) NMDA-receptors mediated activation of the *c-fos* proto-oncogene in a model of brain injury. Neuroscience 35: 273–281

Hertzberg EL, Spray DC (1985) Studies of gap junctions: biochemical analysis and use of antibody probes. In: Bennett MVL, Spray DC (eds) Gap junctions. Cold Spring Harbor, New York, pp 57–65

His W (1890) Die Entwicklung des menschlichenRautenhirns vom Ende des ersten zum Beginn des dritten Monats. I. Verlangertes Mar Abh Kg Sachs Ges Wissensch Math Phys Kl 29: 1–74

Holberg M, Duyckaerts C, Durr A. Cancel G, Gourfinkel-ANI, Damier P. Faucheux B, Trotter Y, Hirsch EC, Agid Y, Brice A (1998) Spinocerebellar ataxia type 7 (SCA7): a neurodegenerative disorder with neuronal intranuclear inclusions. Hum Mol Genet 7: 913–918

Horn AKE, Hoffmann KP (1987) Combined GABA-immunocytochemistry and TMB-HRP histochemistry of pretectal nuclei projecting to the inferior olive in rats, cats and monkeys. Brain Res 409: 133–138

Hsu M, Raine L, Fanger H (1981) Use of the avidin-biotin-peroxidase complex (ABC) in immunoperoxidase techniques. A comparison between ABC and unlabeled antibody (PAP) procedures. J Histochem Cytochem 29: 577–580

Huerta MF, Hashikawa T, Gayoso MJ, Harting JK (1985) The trigemino-olivary projections in the cat: contribution of individual subnuclei. J Comp Neurol 241: 180–190

Ikeda Y, Noda H, Sugita S (1989) Olivocerebellar and cerebello-olivary connections of the oculomotor region of the fastigial nucleus in the macaque monkey. J Comp Neurol 284: 463–488

Ito K, Ishikawa Y, Skinner RD, Mrak RE, Morrison-Bogorad M, Mukawa J, Griffin WS (1997) Lesioning of the inferior olive using a ventral surgical approach. Characterization of temporal and spatial astrocytic responses at the lesion site and in the cerebellum. Mol Chem Neuropathol 31: 245–264

Ito M (1990) A new physiological concept on cerebellum. Rev Neurol (Paris) 146: 564–569

Ito M (1993) Synaptic plasticity in the cerebellar cortex and its role in motor learning. Can J Neurol Sci 20: S70-S74

Ito M, Miyashita Y, Ueki A (1978) Functional localization in the rabbit's inferior olive determined in connection with the vestibulo-ocular reflex. Neurosci Lett 8: 283–287

Janota I (1979) Neuroaxonal dystrophy in the neonate. Acta Neuropathol (Berl) 46: 151–154

Jeneskog T (1987) Termination in posterior and anterior cerebellum of a climbing fibre pathway activated from the nucleus of Darkschewitsch in the cat. Brain Res 412: 185–189

Jones N, Stelz T, Batini C, Caston J (1995) Effects of lesions of the inferior olivary complex in learning of the equilibrium behavior in the young rat during ontogenesis. I. Total lesion of the inferior olive by 3-acetylpyridine. Brain Res 697: 216–224

Kabayashi K, Kurachi M, Gyoubu T, Fukutani Y, Inao G, Nakamura I, Yamaguchi N (1990) Progressive dysphasic dementia with localized cerebral atrophy: report of an autopsy. Clin Neurophathol 9: 254–261

Kanda KI, Sato Y, Ikarashi K, Kawasaki T (1989) Zonal organization of climbing fibre projections to the uvula in the cat. J Comp Neurol 279: 138–148

Keating JG, Thach WT (1995) Nonclock behavior of inferior olive neurons: interspike interval of Purkinje cell complex spike discharge in the awake behaving monkey is random. J Neurophysiol 73: 1329–1340

Keating JG, Thach WT (1997) Non clock signal in the discharge of neurons in the deep cerebellar nuclei. J Neurophysiol 77: 2232–2234

King JS (1976) The synaptic cluster (glomeruli) in the inferior olivary nucleus. J Comp Neurol 165: 387–400

King JS (1980) Synaptic organization of the inferior olivary complex. In: J, de Montigny C, Lamare Y (eds) The inferior olivary nucleus. Anatomy and physiology. Raven Press, New York, pp 1–33

King JS, Ho RH, Burry RW (1984) The distribution and synaptic organization of serotoninergic elements in the inferior olivary complex of the opossum. J Comp Neurol 227: 357–368

King JS, Martin GF, Bowman MH (1975) The direct spinal area of the inferior olivary nucleus: An electron microscopic study. Exp Brain Res 22: 13–24

Kitao Y, Nakamura Y, Okoyama S (1983) An electron microscopy study of the cortico-pretecto-olivary projection in the cat by a combinated degeneration and horseradish peroxidase tracing technique. Brain Res 280: 139–142

Kooy FH (1917) The inferior olive in vertebrates. Folia Neuro-Biol 10: 205–369

Korneliussen HK, Jansen J (1964) The morphogenesis and structure of the inferior olive of cetacea. J Hirnforsch 7: 301–314

Künzle H (1983) Supraspinal cell populations projecting to the cerebellar cortex in the turtle (*Pseudemys scripta elegans*). Exp Brain Res 49: 1–12

Künzle H (1985) Climbing fiber projection to the turtle cerebellum: Longitudinally oriented terminal zones within the basal third of the molecular layer. Neuroscience 14: 159–168

Künzle H, Wiklund L (1982) Identification and distribution of neurons presumed to give rise to cerebellar climbing fibers in turtle. A retrograde axonal flow study using radioactive D-aspartate as a marker. Brain Res 252: 146–150

Lafarga M, Hervás J-P, Crespo D, Villegas J (1980) Ciliated neurons in supraoptic nucleus of rat hypothalamus during neonatal period. Anat Embryol 160: 29–38

Landis DM, Rayne HR, Weinstein LA (1989) Changes in the structure of synaptic junctions during climbing fiber synaptogenesis. Synapse 4: 281–293

Landis DMD, Reese TS (1974a) Arrays of particles in freeze-fractured astrocytic membranes. J Cell Biol 60: 316–320

Landis DMD, Reese TS (1974b) Differences in membrane structure between excitatory and inhibitory synapses in the cerebellar cortex. J Comp Neurol 155: 93–126

Landis DMD, Reese TS (1982) Regional organization of astrocytic membranes in cerebellar cortex. Neuroscience 7: 937–950

Landis DMD, Reese TS, Raviola E (1974) Differences in membrane structure between excitatory and inhibitory components of the reciprocal synapses in the olfactory bulb. J Comp Neurol 155: 67–92

Landis DMD, Weinstein LA, Reese TS (1987) Substructure of the postsynaptic density of Purkinje cell dendritic spines revealed by rapit freezing and etching. Synapse 1: 552–558

Lang EJ, Sugihara I, Llinás R (1996) GABAergic modulation of complex spike activity by the cerebello-olivary pathway in rat. J Neurophysiol 76: 255–275

Lang EJ, Sugihara I, Llinás R (1997) Differential roles of apamin- and charybdotoxin-sensitive K+ conductances in the generation of inferior olive rhythmicity in vivo. J Neurosci 17: 2825–2838

Lang EJ, Sugihara I, Welsh JP, Llinás R (1999) Patters of spontaneous Purkinje cell complex spike activity in the awake rat. J Neurosci 19: 2728–2739

Lau KL, Glover RG, Linkenhoker B, Wylie DRW (1998) Topographical organization of inferior olive cells projecting to translation and rotation zones in the vestibulocerebellum of pigeons. Neuroscience 85: 605–614

Lee W, Mitchell P, Tjian R (1987) Purified transcription factor AP-1 interacts with TPA-inducible enchanter elements. Cell 49: 741–752

Leonard CS, Simpson JI, Graf W (1988) Spatial organization of visual messages of the rabbit's cerebellar flocculus. I. Topology of inferior olive neurons of the dorsal cap of Kooy. J Neurophysiol 60; 2073–2090

Llinás R (1974) Eighteenth Bowdich Lecture. Motor aspects of olivocerebellar control. Physiologist 17: 19 46 (cited by Lang et al. 1996)

Llinás R (1989) Electrophysiological properties of the olivo-cerebellar system. In The olivo-cerebellar system in motor control. Strata R (ed) Exp Brain Res Suppl 17; 201–209

Llinás R (1991) The noncontinuous nature of movement execution. In: Humphrey DR, Freund H-J (eds) Motor contral: Concepts and issues. Wiley, Chichester, UK, pp 223–242

Llinás R, Baker R, Sotelo C (1974) Electrotonic coupling between neurons in the cat inferior olive. J Neurophysiol 37: 560–571

Llinás R, Sasaki K (1989) The functional organization of the olivocerebellar system as examined by multiple Purkinje cell recording. Eur J Neurosci 1: 587–602

Llinás R, Volkind RA (1973) The olivocerebellar system functional properties as revealed by harmaline induced tremor. Exp Brain Res 18: 69–87

Llinás R, Welsh JP (1993) On the cerebellum and motor learning. Curr Opin Neurobiol 3: 958–965

Llinás R, Yarom Y (1981a) Electrophysiology of mammalian inferior olivary neurons in vitro. Different types of voltage-dependent ionic conductances. J Physiol 315: 549–567

Llinás R, Yarom Y (1981b) Properties and distribution of ionic conductances generating electroresponsiveness of mammalian inferior olivary neurons in vitro. J Physiol 315: 569–584

Llinás R, Yarom Y (1986) Oscillatory properties of guinea pig inferior olivary neurons and their pharmacological modulation: An in vitro study. J Physiol 376: 163–183

Lopes-Raman A, Armengol JA (1996) Ipsilateral located olivocerebellar projection neurons of the chick. Neurosci Res 25: 33–40

Lui F, Benassi C, Biral G, Corazza R (1999) Olivofloccular circuit in oculomotor control: binocular optokinetic stimulation. Exp Brain Res 125: 211–216

Mack A, Wolburg H (1986) Heterogeneity of glial membranes in the rat olfactory system as revealed by freeze-fracturing. Neurosci Lett 65: 17–22

Maekawa K, Takeda T (1979) Origin of descending afferents to the rostral parts of dorsal cap of inferior olive which transfers contralateral optic activities to the flocculus. A horseradish peroxidase study. Brain Res 172: 393–405

Marani E (1982) Topographic enzyme histochemistry of the mammalian cerebellum. 5'-nucleotidase and acetylcholinesterase. Thesis. University of Leiden.

Marani E (1986) Topographic histochemistry of the cerebellum. Progr Histichem Cytochem 16. Gustau Fischer Verlag, Stuttgart, New York, pp 1–196

Marani E, Voogd J, Boekee A (1977) Acetylcholinesterase staining in subdivisions of the cat's inferior olive. J Comp Neurol 174: 209–226

Marcos P, Covanas R, Narvaez JA, Tramu G, Aguirre JA, Gonzalez-Baron S (1994) Distribution of gastrin-releasing peptide/bombesin-like immunoreactive cell bodies and fibres in the brainstem of the cat. Neuropeptides 26: 93–101

Mareschal P (1934) L'olive bulbaire; anatomie-ontogènése-phylogènése-physiologie et physiopathologie. Doin, Paris, pp 216

Marr D (1969) A theory of cerebellar cortex. J Phisiol (Lond.) 202: 437–470

Martin GF, Beattie MS, Hughes HC, Linauts M, Panneton M (1977) The organization of reticulo-olivo-cerebellar circuits in the North American opossum. J Comp Neurol 160: 507–534

Martin GF, Dom R, King JS, Robards M, Watson CRR (1975) The inferior olivary nucleus of opossum (Didelphis marsupialis virginiana) its organization and connections. J Comp Neurol 160: 507–534

Martin GF, Henkel CK, King JS (1976) Cerebello-olivary fibers: Their origin, course and distribution in the North American opossum. Exp Brain Res 24: 219–236

Massa PT, Mugnaini E (1982) Cell-cell junctional interactions and characteristic plasma membrane features of cultured rat glial cells. Neuroscience 14: 695–709

Massa PT, Mugnaini E (1985) Cell junctions and intramembrane particles of astrocytes and oligodendrocytes: a freeze-fracture study. Neuroscience 7: 523–538

Massa PT, Szuchet S, Mugnaini E (1984) Cell-cell interactions of isolated and cultured oligodendrocytes: formation of linear occluding junctions and expression of peculiar intramembrane particles. J Neurosci 4: 3128–3139

McGaslin PP, Morgan WW (1987) Activity-induced elevation of cerebellar cyclic GMP occurs in the absence of climbing fiber pathways. Brain Res 414: 381–384

McGrane MK, Eriksson MA, Burne RA, Woodward DJ (1977) The inferior olivary complex in the rat: Gross nuclear organization and topography of olivocerebellar projections. Soc Neurosci Abst 3: 59

Miller TM, Heuser JE (1984) Endocytosis of synaptic vesicle membrane at the frog neuromuscular junction. J Cell Biol 98: 685–698

Mizuno N, Mochizuki K, Akimoto C, Matsushima R (1973) Pretectal projections to the inferior olive in the rabbit. Exp Neurol 39: 498–506

Mlonyeni M (1973) The number of Purkinje cells and inferior olivary neurones in the cat. J Comp Neurol 147: 1–10

Moatamed F (1966) Cell frequencies in the human olivary nuclear complex. J Comp Neurol 128: 109–116

Molinari HH (1984) Ascending somatosensory projections to the dorsal accessory olive: An anatomical study in cats. J Comp Neurol 223: 110–123

Molinari HH (1985) Ascending somatosensory projections to the medial accessory portion of the inferior olive: A retrograde study in cats. J Comp Neurol 232: 523–533

Molinari HH (1987) Ultrastructure of the gracile nucleus projection to the dorsal accessory subdivision of the cat inferior olive. Exp Brain Res 66: 175–184

Molinari HH, Starr KA (1989) Spino-olivary termination on spines in cat medial accessory olive. J Comp Neurol 288: 254–262

Mugnaini E (1969) Ultrastructural studies on the cerebellar histogenesis. II. Maturation of nerve cell populations and establishment of synaptic connections in the cerebellar cortex of the chick.. Llinás

R (ed) Neurobiology of cerebellar evolution and development. Chicago, American medical association, pp 749–782

Mugnaini E (1986) Cell junctions of astrocytes, ependyma and related cells in the mammalian central nervous system, with emphasis on the hypothesis of a generalized functional syncytium of supporting cells. In: Fedoroff S, Vernadakis A (eds). Astrocytes, development, morphology and regional specialization of astrocytes. v 1. Academic Press, pp 327–371

Mugnaini E, Oertel WH (1985) Atlas of the distribution of GABAergic neurons and terminals in the rat CNS. In: Brjörklund A, Hökfelt K (eds). GABA and Neuropeptides in the CNS. Part I. Elsevier, Amsterdam, pp 571–573

Mugnaini E, Walberg F (1964) Ultrastructure of neuroglia. In: Reviews of anatomy and cell biology. Springer-Verlag. Berlin-Gottingen Heidelberg

Musil LS, Goodenough DA (1993) Multisubunit assembly of an integral plasma membrane channel protein, gap junction connexin 43, occurs after exit from the ER. Cell 74: 1065–1077

Nagi JI, Yamamoto T, Hertzberg EL (1989) Immunohistochemical localization of the 27 kDa and 43 kDa gap junction proteins in the central nervous system. Soc Neurosci Abst 15: 276.2

Nagi JI, Yamamoto T, Shiosaka S, Dewar KM, Whittaker ME, Hertzberg EL (1988) Immunohistochemical localization of gap junction proteins in the CNS: a preliminary account. In: Modern cell biology.7. Hertzberg EL. Johnson RG (eds) Alan R Liss Inc, New York, pp 375–389

Nakamura Y, Kitao Y, Okoyama S (1983) Cortico-Darkschewitsch-olivary projection in the cat: an electron microscope study with the aid of horseradish peroxidase tracing technique. Brain Res 274: 140–143

Nelson B, Adams JC, Barmack NH, Mugnaini E (1989) Comparative study of glutamate decarboxylase immunoreactive boutons in the mammalian inferior olive. J Comp Neurol 286: 514–539

Nelson B, Barmack NH, Mugnaini E (1984) A GABAergic cerebellar-olivary projection in the rat. Soc Neurosci Abstr 10:161.7

Nelson B, Barmack NH, Mugnaini E (1986) GABAergic projection from vestibular nuclei to rat inferior olive. Soc Neurosci Abstr 12: 255

Nelson B, Mugnaini E (1988) The rat inferior olive as seen with immunostaining for glutamamic decarboxylase. Anat Embryol 179: 109–127

Nelson B, Mugnaini E (1989) Origin of GABAergic inputs to the inferior olive. In: The olivocerebellar system in motor control. Strata P (ed). Exp Brain Res Series 17. Springer-Verlag, Berlin, pp 86–107

Nemecek S, Wolff J (1969) Light and electron microscopic evidence of complex synapses (glomeruli) in oliva inferior (cat). Experientia 25: 634–636

Nieuwenhuys R, Ten Donkelaar HJ, Nicholson (1998) The central nervous system of vertebrates. Smeets WJAJ, Wicht H (ed). Springer-Verlag, Berlin, pp 997–1748

NieuwenhuysR, Oey P (1983) Topological analysis of the brainstem of the reed fish *Erpetoichthys calaboricus.* J Comp Neurol 213: 220–232

Nunes-Cardozo BN, Van der Want J (1990) Ultrastructural organization of the retino-pretecto-olivary pathway in the rabbit: A combined WGA-HRP tracing and GABA immunocytochemical study. J Comp Neurol 291: 313–327

Ockleford CD, Nevard CHF, Indans I (1987) Structure and function of the nematosome. J Cell Sci 87: 27–44

Olszewski J, Baxter D (1954) Cytoarchitecture of the human brain stem. Karger, Basel

Onodera S (1984) Olivary projections from the mesodiencephalic structures in the cat studied by means of axonal transport of horseradish peroxidase and tritiated amino acids. J Comp Neurol 227: 37–49

Oscarsson O (1969) Termination and functional organization of the dorsal spino-olivocerebellar path. J Physiol 196: 453–478

Oscarsson O (1980) Functional organization of olivary projection to the cerebellar anterior lobe. In: Courville J, de Montigny C, Lamare Y (eds). The inferior olivary nucleus. Anatomy and physiology. Raven Press, New York, pp 279–289

Ovtscharoff W, Eusterschulte B, Zienecker R, Reisert I, Pilgrim C (1992) Sex differences in densities of dopaminergic fibers and GABAergic neurons in the prenatal rat striatum. J Comp Neurol 323: 299–304

Palay SL, Chan-Palay V (1974) Cerebellar cortex. Cytology and organization. Springer Verlag, Berlin

Papez JW (1929) Comparative neurology. In: ThomasY (ed), Crowell Co, New York, pp 208–217

Paracchia C (1980) Structural correlates of gap junction permeation. Int Rev Cytol 66: 81–146

Peinado A, Yuste R, Katz LC (1993) Extensive dye coupling between rat neurocortical neurons during the period of circuit formation. Neuron 10: 103–114

Pellionisz A, Llinás R (1980) Tensorial approach to the geometry of brain function. Cerebellar coordination via metric tensor. Neuroscience 15: 933–946

Perez de la Mora M, Possani LD, Tapia R, Teran L, Palacios R, Fuxe K, Hökfelt T, Ljundahl A (1981) Demonstration of central γ-amino-butyrate-containing nerve terminals by means of antibodies against glutamate decarboxylase. Neuroscience 6: 875–895

Peters A, Palay SL, Webster HF (1976) The fine structure of the nervous. In: The cell and their processes. Second edition. Philadelphia. WB Saunders

Pfenninger K, Akert K, Moor H, Sandri C (1972) The fine structure of freeze-fractured presynaptic membranes. J Neurocytol 1: 129–149

Pilgrim C, Hutchison J (1994) Developmental regulation of sex differences in the brain: can the role of gonadal steroids be redefined? Neuroscience 60: 843–855

Pumplin DW, Reese TS, Llinás R (1981) Are the presynaptic membrane particles the calcium channels? Proc Natl Acad Sci USA 78: 7210–7213

Ramón y Cajal S (1906) Notas preventivas sobre la degeneración y regeneración de la las vias nerviosas centrales. Trab Labor Invest Biol Madrit: 295–301

Ramón y Cajal S (1909) Histologie du Système Nerveux de l'Homme et des Verténrés. v 1. Paris. Moloine

Ramón y Cajal S (1911) Histologie du Systéme Nerveux de l'Homme et des Vertébres. vol II, Maloine, Paris

Ramón y Cajal (1988) Sobre las fibras nervosas de la capa molecular del cerebelo. Rev trimestr Histol 2: 33–41

Robertson HA (1992) Immediate-early genes, neuronal plasticity, and memory. Biochem Cell Biol 70: 729–737

Robertson LT, Stotler WA (1974) The structure and connections of the developing inferior olivary nucleus of the rhesus monkey (*Macaca mulatta*). J Comp Neurol 158: 167–190

Rondi-Reig L, Delhaye-Bouchaud N, Mariani J, Caston L (1997) Role of the inferior olivary complex in motor skills and motor learning in the adult rat. Neuroscience 77: 955–963

Ruigrok TJ (1997) Cerebellar nuclei. The olivary connection. Prog Brain Res 114; 167–192

Ruigrok TJ, van der Burg H, Sabel-Goedknegt E (1996) Locomotion coincides with c-Fos expression in related areas of inferior olive and cerebellar nuclei in the rat. Neurosci Lett 214: 119–122

Ruigrok TJH, De Zeeuw CI (1993) Electron microscopy of in vivo recorded and intracellularly injected inferior olivary neurons and their GABAergic innervation in the cat. Microsc Res Tech 24: 85–102

Ruigrok TJH, De Zeeuw CI, van der Burg H, Voogd J (1990a) Intracellular labeling of neurons in the medial accessory olive in the cat. I. Physiology and light microscopy. J Comp Neurol 300, 462–477

Ruigrok TJH, de Zeeuw CI, Voogd J (1990b) Hypertrophy of the inferior olivary neurons: A degenerative, regenerative or plasticity phenomenon. Eur J Morphol 28: 224–239

Ruigrok TJH, Osse RJ, Voogd J (1992) Organization of inferior olivary projections to the flocculus and ventral paraflocculus of the rat cerebellum. J Comp Neurol 316: 129–150

Ruigrok TJH, Voogd J (1988) Evidence for cerebello-midbrain-olivary circuits in rat using PHA-L anterograde and bold labeled WGA-BSA retrograde tracing. Eur J Neurosci (Suppl) 10: 3

Ruigrok TJH, Voogd J (1990) Cerebellar nucleo-olivary projections in the rat: An anterograde tracing study with *Phaseolus vulgaris*-Leocoagglutinin (PHA-L). J Comp Neurol 298: 315–333

Ruigrok TJH, Voogd J (1995) Cerebellar influence on olivary excitability in the cat. Eur J Neurosci 7: 679–693

Rutherford JG, Gwyn DG (1977) Gap junction in the inferior olivary nucleus of the squirrel monkey, *Saimiri sciureus*. Brain Res 128: 347–378

Rutherford JG, Gwyn DG (1980) A light and electron microscopic study of the inferior olivary nucleus of the squirrel monkey, *Saimiri sciureus*. J Comp Neurol 189: 127–157

Saez JC, Berthoud VM, Moreno AP, Spray DC (1993) Gap junctions: multiplicity of controls in differentiated and undifferentiated cells and possible functional implications. In: Shenolikar S, Nairn A (eds.) Advances in Second Messenger and Phosphoprotein Research. v 27. Raven, New York, pp.163–198

Sagar SM, Sharp FR, Curran T (1988) Expression of c-fos protein in brain: a novel method of neuroanatomic metabolitic mapping at cellular level. Science (Washington DC) 240: 1328–1331

Saigal RP, Karamanlidis AN, Voogd J, Michaloudi H, Mangana O (1983) Olivocerebellar connections in sheep studied with the retrograde transport of horseradish peroxidase. J Comp Neurol 217: 440–448

Saint-Cyr JA (1983) The projection from the motor cortex to the inferior olive in the cat: An experimental study using axonal transport techniques. Neuroscience 10: 667–684

Saint-Cyr JA (1987) Anatomical organization of cortico-mesencephalo-olivary pathways in the cat as demonstrated by axonal transport techniques. J Comp Neurol 257: 39–59

Saint-Cyr JA, Courville J (1980) Projections from motor cortex, midbrain and vestibular nuclei to the inferior olive in the cat: Anatomical organization and functional correlates. In: Courville J, de Montigny C, Lamare Y (eds) The inferior olivary nucleus. Anatomy and physiology. Raven Press, New York, pp 97–124

Saint-Cyr JA, Courville J (1982) Descending projections to the inferior olive from the mesencephalon and superior colliculus in the cat. Exp Brain Res 45: 333–345

Saint-Gyr JA, Courville J (1979) Projection from the vestibular nuclei to the inferior olive in the cat. An autoradiographic and horseradish peroxidase study. Brain Res 165: 189–200

Sandri C, van Buren JM, Akert K (1977) Membrane morphology of the vertebrate nervous system: a study with freeze-etch technique. Prog Brain Res 46

Sasaki K, Bower JM, Llinás R (1989) Multiple Purkinje cell recording in rodent cerebellar cortex. Eur J Neurosci 1: 572–586

Sasaki K, Llinás R (1985) Evidence for dynamic electrotonic coupling in mammalian inferior olive in vivo. Soc Neurosci Abstr 11: 181

Scheibel ME, Scheibel AB (1955) The inferior olive. A Golgi study. J Comp Neurol 102: 77–132

Scheibel ME, Scheibel AB, Walber F, Brodal A (1956) Areal distribution of axonal and dendritic patterns in inferior olive. J Comp Neurol 106: 21–49

Schild RF (1970) On the inferior olive of the albino rat. J Comp Neurol 140: 255–260

Schnapp B, Mugnaini E (1975) The myelin sheath: Electron microscopic studies with thin sections and freeze-fracture. In: Colgi Central Symposium: Perspectives in Neurobiology. Santini M (ed) Raven Press, New York, pp 209–233

Shanklin WM (1930) The central nervous system of *Chameleon vulgaris*. Acta Zool 14: 425–491

Sjölund B, Björklund A, Wiklund L (1977) The indolaminergic innervation of the inferior olive. 2. Relation to harmaline induced tremor. Brain Res 131: 23–37

Sjölund B, Wiklund L, Björklund A (1980) Functional role of serotoninergic innervation of inferior olivary cells. In: Courville J, de Montigny C, Lamare Y (eds) The inferior olivary nucleus. Anatomy and physiology. Raven Press, New York, pp 163–169

Soasa-Pinto A, Brodal A (1969) Demonstration of a somatotopical pattern in the cortico-olivary projection in the cat. An experimental anatomical study. Exp Brain Res 8: 364–386

Sotelo C, Arsenio-Nunes ML (1976) Development of Purkinje cells in absence of climbing fibers. Brain Res 111: 389–395

Sotelo C, Bourrat F, Triller A (1984) Postnatal development of the inferior olivary complex in the rat. II. Topographic organization of the immature olivocerebellar projection. J Comp Neurol 222: 177–199

Sotelo C, Chedotal A (1997) Development of the olivocerebellar projections. Perspect Dev Neurobiol 5: 57–67

Sotelo C, Gotow T, Wassef M (1986) Localization of glutamic-acid-decarboxylase-immunoreactive axon terminals in the inferior olive of the rat, with special emphasis on anatomical relations between GABAergic synapses and dendrodendritic gap junctions. J Comp Neurol 252: 32–50

Sotelo C, Llinás R, Baker R (1974) Structural study of inferior olivary nucleus of the cat: Morphological correlates of electrotonic coupling. J Neurophysiol 37: 541–559

Sotelo C, Palay SL (1968) The fine structure of the vestibular nucleus of the rat. J Cell Biol 36: 151–179

Sotelo C, Palay SL (1971) Altered axons and terminals in the lateral vestibular nucleus of the rat. Possible example of axonal remodeling. Lab Invest 25: 653–671

Sotelo C, Wassef M (1991) Cerebellar development: afferent organization and Purkinje cell heterogeneity. Philos Trans R Soc Lond B Biol Sci 331: 307–313

Sotelo, C. and Llinás R (1972) Specialized membrane junctions between neurons in the vertebrate cerebellar cortex. J. Cell Biol 53: 271–289.

Spacek J (1982) 'Free' postsynaptic-like densities in normal adult brain: their occurrence, distribution, structure and association with subsurface cisterns. J Neurocytol 11: 693–706

Spence SJ, Saint-Cyr JA (1988) Comparative topography of projections from the mesodiencephalic junction to the inferior olive, vestibular nuclei, and upper cervical cord in the cat. J Comp Neurol 268: 357–374

Stefanova N, Bozhilova-Pastirova A, Ovtscharoff W (1997a) Distribution of GABA-immunoreactive nerve cells in the bed nucleus of the stria terminalis in male and female rats. Eur J Histochem 41: 23–28

Stefanova N, Bozhilova-Pastirova A, Ovtscharoff W (1997b) Sex differences of parvalbumin-immunoreactive neurons in some rat brain areas. Biomed Rev 7: 91–96

Stensaas LJ, Stensaas SS (1968) Light microscopy of glial cells in turtles and birds. Z Zellforsch 91: 315–340

Sternberger LA (1979) Immunocytochemistry. 2nd edn. John Wiley and Sons, New York

Strata P (1984) Inferior olive: Functional aspects. In: Bloedel JR, Dichans J, Precht W (eds) Cerebellar functions. Springer Verlag, Berlin, pp 230–246

Strata P, Montarolo PG (1982) Functional aspects of the inferior olive. Arch Ital Biol 120: 321–329

Strata P, Rossi F (1998) Plasticity of the olivocerebellar pathway. Trends Neurosci 21: 407–413

Suigihara I, Lang EJ, Llinás R (1995) Serotonin modulation of inferior olivary oscillations and synchronicity: a multiple-electrode study in the rat cerebellum. Eur J Neurosci 7: 521–534

Surchev L (1992) Freeze-etched postsynaptic membranes in the visual cortex reveal different types of synapses including mixed synapses. Brain Res 573: 174–178

Suzuki T, Abe-Dohmea S, Tanaka T (1992) P400 protein is one of the major substrates for Ca2+/calmodulin-dependent kinase II in the postsynaptic density-enriched fraction isolated from rat cerebral cortex, hippocampus and cerebellum. Neurochem Int 20: 61–67

Swenson RS, Castro AJ (1983) The afferent connections of the inferior olivary complex in rats. An anterograde study using autoradiographic and axonal degeneration techniques. Neuroscience 8: 259–275

Szentágothai J, Rajkovits K (1959) Ueber den Ursprung der Kletterfasern des Kleinhirns. Z Anat Entwicklungsgeschichte 121: 130–141

Taber E (1961) The cytoarchitecture of the brain stem of the cat. I. Brain stem nuclei of cat. J Comp Neurol 116: 27–69

Takeda T, Maekawa K (1989) Olivary branching projections to the flocculus, nodulus and uvula in the rabbit. II. Retrograde double labelling study with fluorescent dyes. Exp Brain Res 76: 323–332

Takeuchi Y, Sano Y (1983) Immunohistochemical demonstration of serotonin-containing nerve fibers in the inferior olivary complex of the rat, cat and monkey. Cell Tiss Res 231: 17–27

Tan J, Gerrits NM, Nanhoe R, Simpson JI, Voogd J (1995) Zonal organization of the climbing fiber projection to the flocculus and nodulus of the rabbit: A combined axonal tracing and acetylcholinesterase histochemical study. J Comp Neurol 356: 23–50

Tao-Cheng JH, Bressler JP, Brightman MW (1992) Astroglial membrane structure is affected by agents that raise cyclic AMP and by phosphatidylcholine phospholipase C. J Neurocytol 21: 458–467

Tolbert DL, Massopust LC, Murphy MG, Young PA (1976) The anatomical organization of the cerebello-olivary projection in the cat. J Comp Neurol 170: 525–544

Tweedle CD, Hatton GI (1986) Vacant postsynaptic densities on supraoptic dendrites of adult rats diminish in number with chronic stimuli. Cell Tiss Res 245: 37–41

Ueyama T, Houtani T, Nakagava H, Baba K, Ikeda M, Yamashita T, Sugimoto T (1994) A subpopulation of olivocerebellar projection neurons express neuropeptide Y. Brain Res 634: 353–357

Uylings HBM, van Eden CG, Hofman MA (1986) Morphometry of size-volume variables and comparison of the bivariate relations in the nervous system under various conditions. J Neurosci Methods 18: 19–37

Van der Linde JAM, ten Donkelaar HJ (1987) Observations on the development of cerebellar afferents in *Xenopus leavis*. Anat Embryol 176: 431–439

Van der Linde JAM, ten Donkelaar HJ, de Boer-van Huizen (1990) Development of olivocerebellar fibers in the clawed toad, *Xenopus leavis*: A light and electron microscopical HRP study. J Comp Neurol 293: 236–252

Van der Togt C, Van der Want (1992) Variation in form and axonal termination in the nucleus of optic tract of the rat: The medial terminal nucleus input on neurons projecting to the inferior olive. J Comp Neurol 325: 446–461

Van der Want JJL, Voogd J (1987) Ultrastructural identification and localization of climbing fiber terminals in the fastigial nucleus of the cat. J Comp Neurol 258: 81–90

Van der Want JJL, Wiklund L, Guegan M Ruigrok T, Voogd J (1989) Anterograde tracing of the rat olivocerebellar system with *Phaseolus vulgaris*-leucoagglutinin (PHA-L). Demonstration of climbing fiber collateral innervation of the cerebellar nuclei. J Comp Neurol 288: 1–18

Van Houten M, Brawer J (1979) Regional variation in glia and neurons within the hypothalamic ventromedial nucleus. J Comp Neurol 179: 719–738

Verhaart WJC, Voogd J (1962) Hypertrophy of the inferior olives in the cat. J Neuroph Exp Neurol 21: 92–104

Vincenzi L (1886–1887) Sulla fina anatomia dell oliva bulbare dell'usmo. Estr della Real Accad Medic di Roma II. 3: 171–182

Vogt-Nelsen L (1954) The inferior olive in birds. A comparative morphological study. J Comp Neurol 101: 447–481

Voogd J (1982) The olivocerebellar projection in the cat. In: Palay SL, Chan-Palay V (eds): The cerebellum-New Vistas. Berlin, Heidelberg, New York, Springer-Verlag. Exp Brain Res Series 6, pp 134–161

Voogd J (1989) Parasagittal zones and compartments of the anterior vermis of the cat cerebellum. In: Strata P (ed). The inferior olive in the motor control. Springer-Verlag, Berlin, pp 3–19

Voogd J, Bigarè F (1980) Topographical distribution of olivary and corticonuclear fibers in the cerebellum. In: Courville J, de Montigny C, Lamare Y (eds) The inferior olivary nucleus. Anatomy and physiology. Raven Press, New York, pp 207–305

Voogd J, Eisenman LM, Ruigrok TJH (1993) Relation of olivocerebellar projection zones to zebrin pattern in rat cerebellum. Soc Neurosci Abstr 19: 1216

Walberg F (1980) Olivocerebellocortical projection in the cat as determined with the method of retrograde axonal transport of horseradish peroxidase. I. Topographical pattern. In: Courville J, de Montigny C, Lamare Y (eds) The inferior olivary nucleus. Anatomy and physiology. Raven Press, New York, pp 169–186

Walberg F (1956) Descending connections to the inferior olive: A experimental study in the cat. J Comp Neurol 104: 77–173

Walberg F (1963) An electron microscopic study of the inferior olive of the cat. J Comp Neurol 120: 1–18

Walberg F (1964) Further electron microscopical investigations of the inferior olive of the cat. In: Bergmann W, Schade JP (eds). Topic in basic neurology. Prog Brain Res VI, Elsevier, Amsterdam, pp 59–75

Walberg F (1974) Descending connections from the mesencephalon to the inferior olive: An experimental study in the cat. Exp Brain Res 21: 145–156

Walberg F (1982) Olivary afferents from the brain stem reticular formation. Brain Res 471: 130–136

Walberg F, Ottersen O (1989) Demonstration of GABA immunoreactive cells in inferior olive of baboons (*Papio papio* and *Papio anubis*). Neurosci Lett 101: 149–155

Walberg F, Pompeiano O, Brodal A, Jansen J (1962) The fastrigiovestibular projections in the cat. An experimental study with silver impregnation methods. J Comp Neurol 118: 49–75

Wassef M, Chedotal A, Cholley B, Thomasset M, Heizmann CW, Sotelo C (1992a) Development of the olivocerebellar projection in the rat: I. Transient biochemical compartmentation of the inferior olive. J Comp Neurol 323: 519–536

Wassef M, Cholley B, Thomasset M, Heizmann CW, Sotelo C (1992b) Development of the olivocerebellar projection in the rat: II. Matching of the developmental compartmentations of the cerebellum and inferior olive through the projection map. J Comp Neurol 323: 537–550

Waxman SG, Black JA (1984) Freeze-fracture ultrastructure of the perinodal astrocytes and associated glial junctions. Brain Res 308: 77–87

Welsh JP, Chang B, Menaker ME, Aicher SA (1998) Removal of the inferior olive abolishes myoclonic seizures associated with a loss of olivary serotonin. Neuroscience 82: 879–897

Welsh JP, Lang EJ, Sugihara I, Llinás R (1995) Dynamic organization of motor control within the olivocerebellar system. Nature 374: 453–457

Wentzel PR, Wylie DR, Ruigrok TJ, De Zeeuw CI (1995) Olivary projecting neurons in the nucleus prepositus hypoglossi, group y and ventral dentate nucleus do not project to the oculomotor complex in the rabbit and the rat. Neurosci Lett 190: 45–48

Westrum LE (1980) Alterations in axons and synapses of olfactory cortex following olfactory bulb lesions in newborn rats. Anat Embryol 160: 153–172

Wharton SM, Payne JN (1985) Axonal branching in parasagittal zones of the rat olivocerebellar projection: a retrograde fluorescent double labelling study. Exp Brain Res 58: 183–189

Whitlock DG (1952) A neurohistological and neurophysiological study of afferent fiber tracts and receptive areas of avian cerebellum. J Comp Neurol 97: 567–635

Whitworth RH Jr, Haines DE (1986a) On the question of nomenclature of homologous subdivisions of the inferior olivary complex. Arch Ital Biol 124: 271–317

Whitworth RH, Haines DE (1986b) The inferior olive of *Saimiri sciureus*: Olivocerebellar projections to the anterior lobe. Brain Res 372: 55–71

Whitworth RH, Haines DE, Patric GW (1983) The inferior olive of a prosimian primate, *Galago senegalensis*. II. Olivocerebellar projections to the vestibulocerebellum. J Comp Neurol 219: 228–240

Whitworth RH, Haines DE, Patric GW (1984) Olivocerebellar projections to paramedian lobule in tree shrew (*Tupaia glis*): a horseradish peroxidase study. Brain Res 305: 271–282

Wiklund L, Descarries L, Møllgard K (1981a) Serotoninergic axon terminals in the rat dorsal accessory olive: normal ultrastructure and light microscopic demonstration of regeneration after 5,6-dihydroxytryptamine lesioning. J Neurocytol 10: 1009–1027

Wiklund L, Sjölund B, Björklund A (1981b) Morphological and functional studies on the serotoninergic innervation of the inferior olive. J Physiol (Paris) 77: 183–186

Wiklund L, Toggenburger G, Cuénod M (1982) Aspartate: possible neurotransmitter in cerebellar climbing fibers. Science 216: 78–80

Wiklund L, ToggenburgerG, Cuénod M (1984) Selective retrograde labelling of the rat olivocerebellar climbing fiber system with D-[3H]aspartate. Neuroscience 13: 441–468

Wilczynski W (1982) Brainstem afferents to the cerebellum in the leopard frog *Rana pipiens*. Anat Rec 202: 203 A

Williams EM (1909) Vergleichend anatomosche Studien über den Bau und die Bedeutung der Oliva inferior der Säugetiere und Vögel. Arb Neur Inst Wiener Univ 17: 118–149

Willis T (1664) Cerebri anatome, cui accessit nervorum discriptio et usus.v 2 English translation 1965. Feindel W (ed). McGill University press, Montreal, p 146 (cited by Bowman and King 1973)

Wujek JR, Reier PJ (1984) Astrocytic membrane morphology: differences between mammalian and amphibian astrocytes after axotomy. J Comp Neurol 222: 607–619

Yagishita S, Itoh Y, Nacano T (1986) Hypertrophy of the olivary nucleus. An ultrastructural study. Acta Neuropathol (Berl) 69: 132–138

Yoshimura K (1909) Experimentalle und vergleichend anatomische Untersuchungen über die undere Olive der Vögel. Arb Neur Inst Wiener Univ 18: 46–59

Zilles K (1992) Neuronal plasticity as an adaptive property of the central nervous system. Ann Anat 174: 383–391

Subjext Index